Driving Ultimate Project Performance®

Transforming from Project Manager to Project Leader

- BUILD A TRANSFORMATION FOUNDATION
- EMBRACE THE 7-HABITS OF SUCCESSFUL PROJECT LEADERS
- HARNESS THE WISDOM OF OTHERS
- FINALIZE AND EXECUTE YOUR PLAN

A Self-Paced Leadership Development Program

Includes 30 Developmental Exercises • Online Coaching
Virtual Evaluations & Feedback • Web-based Learning

Workbook and Coaching Guide

Ervin (Earl) Cobb, MSEE   Jim Grigsby, CRCE, CHCS

Published by
RICHER Press
2320 East Baseline Road, Suite 148-214
Phoenix, Arizona 85042

Published in the United States with printing and distribution in United States and the United Kingdom.

All rights reserved. No Part of this book may be Reprinted or reproduced or utilized in any form or by Electronic, mechanical or other means including Photocopying and recording, or in any information Storage of retrieval system, without permission in writing from the publisher.

©Richer Life, LLC

DRIVING ULTIMATE PROJECT PERFORMANCE
Transforming from Project Manager to Project Leader

Ervin (Earl) Cobb and Jim Grigsby

Edited by Charlotte D. Grant-Cobb, PhD, PCC

ISBN: 978-0-9988773-4-1

Library of Congress Control Number: 2017959760

First Printing March 2018

Cover Design: RICHER Media USA

CID: 57254664X

CONTENTS

GETTING THE MOST FROM THE WORKBOOK

HOW TO USE THE WORKBOOK

INTRODUCTION	7
PART I: BUILDING A TRANSFORMATION FOUNDATION	9
1. Creating a Compelling Vision of You – The Outstanding Project Leader	13
2. Analyzing Your Current Organizational Alignment, Project Leadership Tendencies and Development Perspective	65
3. Planning Your Transformation Journey	79
4. Imagining and Constructing Your Transformation Support Team and Communications Strategy	85
PART II: EMBRACING THE 7-HABITS OF SUCCESSFUL PROJECT LEADERS	99
5. Communicating Vertically, Horizontally and Often	103
6. Planning as a Team, Executing as a Team, Winning as a Team	107
7. Proactively and Fearlessly Managing Project and Resource Change	111
8. Linking Risk to Common Objectives with a Compelling Vision	115
9. Approaching Problem Solving as a Creative and Analytical Process	119
10. Maintaining Project Quality with Incremental Measures and Team Focus	123
11. Accepting Your Role as the Chief Confidence Builder	127
PART III: HARNESSING THE WISDOM OF OTHERS	131
12. Thinking Like a Leader – Perspective and Learning	135
13. Understanding and Using the Power of Micro-Messaging	147
14. Identifying Career Critical Mentor Relationships	153
15. Developing Your "Mentor Ask" and Follow-up Strategy	161
PART IV: DEVELOPING AND SYSTEMATICALLY EXECUTING YOUR ACTION PLAN	169
16. Constructing Your Transformation Health Insight Summary (THIS)	173
17. The Action Planning Process and Determining Your Guidelines	181
18. Developing Your Transformation Action Plan *Via Transformation Harvesting Actions and Thoughts (THAT)*	191
19. Finalizing and Executing Your Action Plan	215
20. Transformation Action Tracking	223
LIST OF WORKBOOK EXERCISES	237
ABOUT THE AUTHORS	239

A Project Leader is a Project Manager, but a Project Manager is not necessarily a Project Leader.

GETTING THE MOST FROM THE WORKBOOK

Before you get started, take a moment to think about why you purchased this workbook. Understanding the transformation process and expectations about the exercises will help you focus on accomplishing the desired result.

In order to help clarify, consider the following questions:

- What are the five to seven events and/or choices that brought you to where you are today professionally and personally?
- How did these events/choices contribute to you choosing to purchase and participate in the DUPP Program?
- What do you hope to gain from your investment in leadership development?
- What meaningful impact will it produce in your professional career and personal life?

In addition to your reflection on the above questions, here are some ideas we recommend to help you get the most out of this Program. It is our experience that people who adhere to the following agreement will have a deeper and more enriching overall experience. By participating in this fashion, you will generate a more candid evaluation of yourself and most effectively take advantage of what this workbook has to offer. Take a moment to reflect on the suggestions below.

AGREEMENT	SUGGESTED ACTIONS OR BEHAVIORS
Be fully present.	Let go of thoughts about other activities while you read. Bring your full attention to the work.
Take responsibility for your own success.	Act as though you are 100% responsible for the outcome of your engagement with this material. Because you are.
Participate as fully as possible.	Complete all the exercises to the best of your abilities. Apply the concepts and skills that work best for you, and modify those that do not.
Practice good life management.	Invest time at scheduled intervals to work on the materials when you are mentally and emotionally at your best.
Lean into your discomfort.	Be candid, open, and direct. Allow yourself to be curious and vulnerable
Take the process seriously, and more importantly take yourself lightly. Make this a positive and rewarding experience.	Allow yourself balance. Find the learning and humor in both your successes and mistakes. Most importantly, have fun!

HOW TO USE THE WORKBOOK

After the introduction to *Driving Ultimate Project Performance®*, each subsequent section builds on a series of discussions, illustrations, exercises and targeted questions designed to guide you through the process of developing your abilities and the mindset required to become an Outstanding Project Leader. We recommend that you use the following sequence to help efficiently digest the material and guidance:

Read Intently

Read each section through completely, as we introduce and illustrate an integrated set of steps designed to build your leadership consciousness, Project Leadership focus and transform your thinking to that of an Outstanding Project Leader.

Contemplate

Embracing a set of carefully chosen principles and specifically designed exercises will help you bring the vision of you, the Outstanding Project Leader, to life. Through a process of dynamic examination and reflection, you will be encouraged to contemplate some significant, real-life implications as well as learn from others.

Be Open to Becoming a Leader and Growing into the Role

As you sequentially build your understanding of how a Leader thinks, you will begin noticing habits and conditioned patterns that present you with clear opportunities for growth as a Project Leader. Though you may encounter personal resistance along the way, you will also discover new and exciting strengths.

Allow Your Transformation to Morph at the Required Pace

As you become more proficient at *leveraging* the wisdom of mentors, *utilizing* the power of micro-messaging and *responding* to both personal and professional situations with your expanded leadership thought, you will began to see the results of your transformation.

INTRODUCTION

DRIVING ULTIMATE PROJECT PERFORMANCE
TRANSFORMING FROM PROJECT MANAGER TO PROJECT LEADER

"Everyone is one good decision away from a better life" and *you* are one-step closer because you made the decision to participate in the *Driving Ultimate Project Performance (DUPP) Program* and to take advantage of the self-paced *DUPP Workbook and Coaching Guide.* Your decision to take action to improve your Project Leadership skills will create a more fulfilling and impactful life. What is next?

For years, we have been teaching Leadership, living Leadership and, unfortunately, observing some "misguided attempts at Leadership". We both rose to leadership roles in fish bowls, guided by mentors and driven by the desire to succeed and make a difference. The fact that you purchased and are planning to make use of the *DUPP Program* indicates you have the same drive and qualities.

Think of the *DUPP* as your personal Leadership framework – the infrastructure components are within the covers, it is incumbent of you to build the structure, paint the interior, add fixtures and decorate it. Our role as author, instructor and coach is to assist you in establishing a solid foundation for this valuable Leadership transformation and to keep you moving forward.

We suggest that you take full advantage of the *DUPP Program* by registering as a Program Participant at our website (www.leadershipchecklist.com). By registering, you will have access to 24/7 email response to your questions, be able to access the Program's FREE Web-based learning tools and receive "real-time," complimentary email feedback by a professional Leadership Development Coach on your progress.

The essence of the *DUPP* Program *is* to build on your desire to move forward, at your own pace. There are opportunities for you to learn about your Leadership style, strengthen your Leadership skills, build a solid foundation for your Leadership transformation and become the Outstanding Project Leader you desire to be.

Embrace the opportunity to build a more fulfilling life and thank you for inviting us along for the ride.

PART I

BUILDING A TRANSFORMATION FOUNDATION

"To succeed, you will soon learn, as I did, the importance of a solid foundation in the basics of education - literacy, both verbal and numerical, and communication skills."

— Alan Greenspan

PART I
BUILDING A TRANSFORMATION FOUNDATION

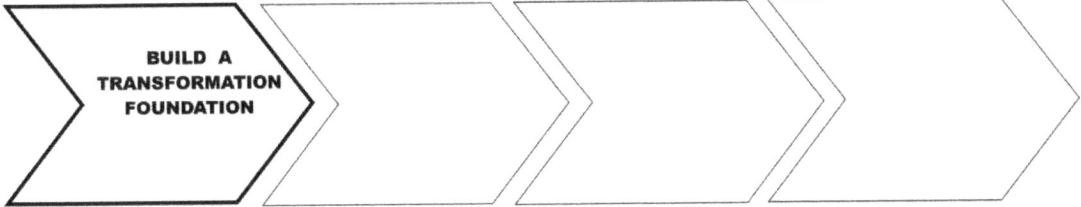

Building a Transformation Foundation is a critical part of being able to transform your thinking from that of a good Project Manager to an Outstanding Project Leader.

Every successful building, institution, organization or team has a strong foundation. There are no exceptions. It is equally important for every Leader to have a strong foundation – as you progress through this workbook, you will begin to build the foundation required to move quickly toward becoming you, the Outstanding Project Leader. The chart below shows the four building blocks involved in the foundation building process.

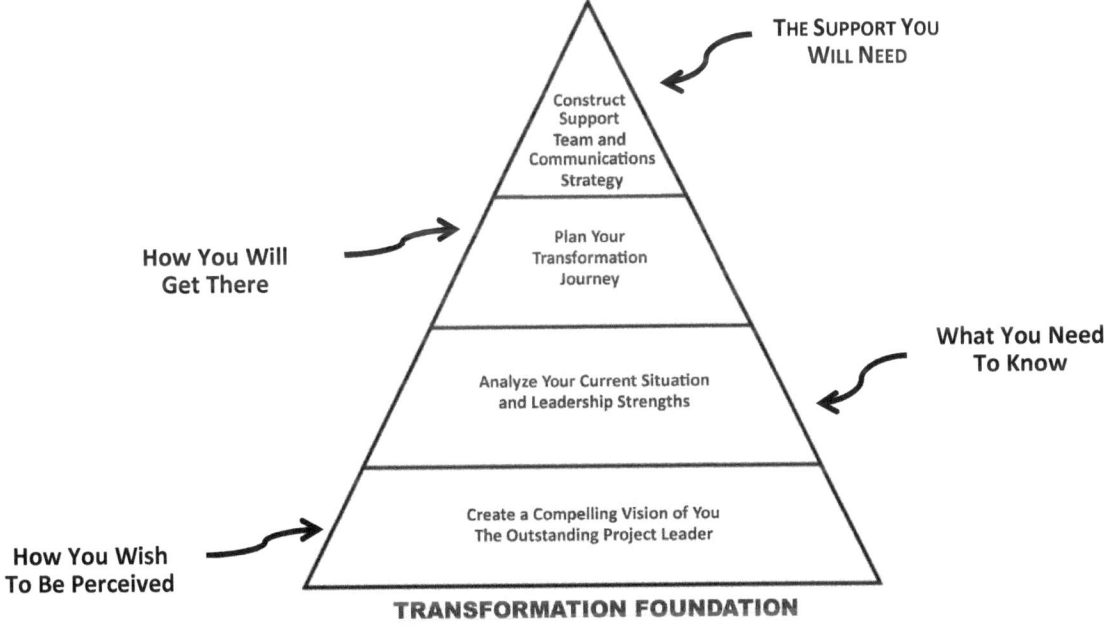

Without carefully establishing each of these building blocks, it will be difficult to transform into an influential and respected Project Leader. Remember this section builds your foundation and your launching pad. It will prepare you to succeed. So, take your time and complete all the exercises. You will use the insights and findings in later sections.

Are you ready to build your transformation foundation? Let's get started!

"Create a compelling vision, one that takes people to a new place, and then translate that vision into a reality."

— Warren G. Bennis

1.

Creating a Compelling Vision of You the Outstanding Project Leader

Before you can seriously believe in a goal or objective, you first must have an idea of what it looks like. To paraphrase the old adage: *"We must see it before we can believe it."* Visualizing the *"Outstanding Project Leader"* in you is simply a technique for creating a mental image of a future "you". When you visualize your desired outcome, you begin to "see" the possibility of achieving it. Through visualization, you catch a glimpse of what can be.

In this section, you are making the first step in the four-step process of building a foundation that will be used to support and guide you through your transformation from being a good *Project Manager* to becoming an *Outstanding Project Leader*.

The comprehensiveness of the exercises in this section coupled with your experience and introspection will give you the insight into yourself, and your current organization, needed to make substantive personal and professional change.

Each exercise in this section differs, but when combined they will help direct you through an exploration that highlights your unique, individual experience while simultaneously considering the groups and organizations to which you belong.

The first step in starting your Leadership development is to cultivate a sense of clarity about "your" vision of "you", the Outstanding Project Leader. This vision will provide the direction and magnitude for your transformation. It will also provide you with both the personal and professional aspiration required to take action. Your vision and aspiration will help you decide where best to invest your time and energy during the transformation process. When your objectives and actions are aligned with your goals, they will drive the impact needed to successfully complete the challenging transformation.

As you move forward in this unique visioning process, you will begin to develop the insights and perspectives required to consistently *"think like a Leader"*.

As a part of the visioning process, it is important to consider the context of your Project Leadership role within your current organization. If you are clear about both the personal and professional aspects of the vision, you can evaluate where and how you will fit within that organization after your transformation.

On the other hand, if your vision differs significantly from what you are currently doing, how you desire to work and the culture within your current organization, the new found revelations can help you identify the "gaps" that exist and determine your options.

In addition to creating a well-defined and compelling vision, it is also important to be clear about your motivation. The combination of vision and motivation is what will enable you to maximize your potential for becoming an Outstanding Project Leader. Without the sufficient motivation, a solid vision and an understanding of your current capabilities and opportunities, you are likely to struggle when progress during the transformation process becomes difficult.

Your future waits. Let's move forward.

Leadership Performance Assessment

The transformation from Manager to Leader requires that you gain important insights into yourself to make substantive personal and professional change. In most cases, the change requires adjustments in the way to see, to treat and to think about others who must be involved in your success.

Although individual strengths and abilities may vary, most experts agree that outstanding Leaders tend to view the world and think in similar ways. A solid set of Leadership Attributes is particularly important, as is a commitment to high ethical standards. Where "gaps" exist, opportunities also exist to improve your ability to easily reflect these attributes and/or compensate for deficiencies, overtime, to yield consistently outstanding results.

This Leadership Assessment consists of ten exercises. Each exercise is designed to help capture your current thoughts and perspectives (your "as-is" state-of-mind) relative to deploying a specific set of *Key Leadership Attributes*. The ten Key Leadership Attributes are listed and defined in the table below.

KEY LEADERSHIP ATTRIBUTE	OUTSTANDING PROJECT LEADERS...
1. Leads by Example	...serve as an example for others.
2. Visionary	...provide a sense of direction for others.
3. Strong Communicator	...listen and consider as well as speak.
4. Trustworthy	...behave in a consistent, highly ethical and fair manner.
5. Calm	...stay cool under pressure, especially when things go wrong.
6. Open-Minded	...invite constructive dissent or disagreements.
7. Professional	...are respectable both at work and in private life.
8. Clear Thinking	...simplify things to maximize understanding and minimize confusion.
9. Nurturing	...show respect for all people, their opinions and their capacity.
10. Supportive	...help others build on their strengths and show appreciation.

You should respond honestly to each of the statements in each exercise in a manner which best reflects how you currently "think" and "see" the world today.

Do not attempt to respond here to what you think might be *best*. At the end of each exercise, you will be asked to tally a score. Your score will be used later in this section to help you perform a gap analysis and determine which gaps are significant enough to possibly prevent a successful transformation at this time and must be addressed.

Once it is determined which gaps can and will be addressed, you will then be in a position to construct your Personal Vision Statement. This Vision will help you visualize yourself as the Outstanding Project Leader you desire to be. It will also aid in the development of personal and professional affirmations you can use to provide daily focus, encouragement and inner support throughout your transformation journey.

LEADS BY EXAMPLE
Serves as an Example for Others

Please respond to each of the statements below based on how you currently "think" and how you "see the world" <u>today</u> by circling the most appropriate number (1-10) along the continuum.

Not at All — Sometimes — Always

1	2	3	4	5	6	7	8	9	10

1. I am sensitive to team member's feelings, and am always kind to them.

1	2	3	4	5	6	7	8	9	10

2. I take time to make team members feel special.

1	2	3	4	5	6	7	8	9	10

3. I listen to each Team Member's emotions as well as words.

1	2	3	4	5	6	7	8	9	10

4. I perceive each Team Member's needs and wants as being valid.

1	2	3	4	5	6	7	8	9	10

5. I choose my battles wisely.

1	2	3	4	5	6	7	8	9	10

6. I respect all Team Member differences.

1	2	3	4	5	6	7	8	9	10

7. I avoid being defensive and placing Team Members on the defensive.

1	2	3	4	5	6	7	8	9	10

8. I give Team Members the benefit of the doubt.

1	2	3	4	5	6	7	8	9	10

9. I resolve interpersonal problems as quickly as possible.

1	2	3	4	5	6	7	8	9	10

10. I treat Team Members the way I would like to be treated.

Score: _____
(The sum of the numbers circled for each statement above)

LEADS BY EXAMPLE
Transformation Framework

1. I am sensitive to team member's feelings and am always kind to them.	Everyone has a rough day, or a day when everything seems to go wrong. Outstanding Leaders are sensitive during these times and supportive of their team.
2. I take time to make team members feel special.	They all want to feel special. Outstanding Leaders take the time to sincerely praise their teams for things that they do.
3. I listen to each Team Member's emotions as well as words.	Remember that 60 to 90 percent of our communication is nonverbal. Outstanding Leaders read their Team Members' body language as well as their emotions.
4. I perceive each Team Member's needs and wants as being valid.	Instead of jumping to the conclusion that a Team Member can survive quite well without the things that *they say are needed*, Outstanding Leaders take the position that the request is valid, and then do everything in their power to respond.
5. I choose my battles wisely.	Outstanding Leaders do not waste their time and energy engaging in fights that have no consequence or that will leave them drained.
6. I respect all Team Member differences.	Each one of us is unique. Outstanding Leaders consider their *Team Members'* differences when they make decisions that affect them.
7. I avoid being defensive and placing Team Members on the defensive.	Outstanding Leaders do not take honest feedback personally--they learn from it and use it to improve. When they provide feedback to their Team Members, they make sure that it is candid, fair, and honest, and helps the Team Member find ways to use it to improve.
8. I give Team Members the benefit of the doubt.	No one goes to work each day wanting to do a terrible job. Outstanding Leaders assume that each one of their Team Members is a good and honest person who wants to do the best job possible.

LEADS BY EXAMPLE
Transformation Framework

9. I resolve interpersonal problems as quickly as possible.	Outstanding Leaders should not let disagreements or hurt feelings fester. They get these issues out into the open and work to resolve them quickly.
10. I treat Team Members the way I would like to be treated.	Outstanding Leaders know how they would like to be treated by others and they do the same for their Team Members.

"The reality is that the only way change comes is when you lead by example."

— Anne Wojcicki

WHAT ARE YOUR THOUGHTS REGARDING YOUR CHALLENGES HERE?

Your thoughts on what you have just read are important. Take the time and be honest.

VISIONARY
Provides a Sense of Direction for Others

Please respond to each of the statements below based on how you currently "think" and how you "see the world" today by circling the most appropriate number (1-10) along the continuum.

Not at All   Sometimes   Always

1	2	3	4	5	6	7	8	9	10

1. I am inspirational and tap into the emotions of others.

1	2	3	4	5	6	7	8	9	10

2. I consider myself as having a sufficient quantity of emotional intelligence.

1	2	3	4	5	6	7	8	9	10

3. I consider myself a very open-minded professional.

1	2	3	4	5	6	7	8	9	10

4. I am complimented often on the robustness of my imagination.

1	2	3	4	5	6	7	8	9	10

5. I consider myself resolute and I hold high convictions in face of adversity.

1	2	3	4	5	6	7	8	9	10

6. I have a strong inner resolve and I am seen as being persistent.

1	2	3	4	5	6	7	8	9	10

7. I approach most team challenges in a collaborative fashion.

1	2	3	4	5	6	7	8	9	10

8. I am often described as being bold and courageous.

1	2	3	4	5	6	7	8	9	10

9. In a magnetic fashion, I always invite team members to make my vision their own.

1	2	3	4	5	6	7	8	9	10

10. I consider myself a very optimistic person.

Score: _____
(The sum of the numbers circled for each statement above)

VISIONARY
Transformation Framework

1. I am inspirational and tap into the emotions of others.	Outstanding Leaders drive emotions of team members in the right direction to bring out the best in them.
2. I consider myself as having a sufficient quantity of emotional intelligence.	Outstanding Leaders inspire with consistency. They are aware of their emotions and are empathic (aware of the feelings of others). Only through empathy, can a leader connect with the hearts of their team and inspire them to realize their greatness.
3. I consider myself a very open-minded professional.	Although visionaries keep a big picture in mind, they are flexible on how they get there. Outstanding Leaders are receptive to new information and can maintain multiple perspectives. This open-mindedness allows them to navigate difficult situations with a flexible mind, pulling from many resources and arriving at creative solutions.
4. I am complimented often on the robustness of my imagination.	Visionary Leaders have a childlike playfulness. Outstanding Leaders value their imagination and allow themselves to dream, exercising their mental focus to see beyond what's in the physical world now. They encourage others to dream big.
5. I consider myself resolute and I hold high convictions in face of adversity.	Obstacles are constant. Realizing the vision will not be easy. With inner resolve, Outstanding Project Leaders push through difficulties and setbacks. Setbacks are not a sign of failure to them; they are mere stopping points on the way to realizing their vision. As such, they are willing to take calculated risks and endure uncertainty. They give assurance and clarity to others when facing uncertainty.

VISIONARY
Transformation Framework

6. I have a strong inner resolve and I am seen as being persistent.	Each one of us is unique. Outstanding Leaders consider their Team Members' differences when they make decisions that affect them. They remain agile enough to pivot and make course corrections, but they always persist.
7. I approach most team challenges in a collaborative fashion.	You cannot bring a business vision to life alone. It takes a talented team that can work together. Outstanding Leaders inspire others to harness their unique gifts and strengths to innovate and find creative solutions.
8. I am often described as being bold and courageous.	Outstanding Leaders know that being bold and courageous means having the willingness to initiate and take action. They are willing to take risks. They do not wait for someone else to do something if a situation has the potential of jeopardizing the project or team performance.
9. In a magnetic fashion, I always invite team members to make my vision their own.	Outstanding Leaders are inclusive, inviting others to make the vision their own. They attract talented people who are passionate about what they do. They create thriving, innovative cultures where individuals have the freedom to create their best work and take pride in their efforts. Visionary Leaders bring out the best in their people.
10. I consider myself a very optimistic person.	Outstanding Leaders hold a positive outlook for the future. They are hopeful they will achieve success. They do not view problems as personal, permanent or pervasive. Instead, they see problems as impersonal, temporary and relate only to the present situation.

"You can't do it unless you can imagine it."

— George Lucas

WHAT ARE YOUR THOUGHTS REGARDING YOUR CHALLENGES HERE?

Your thoughts on what you have just read are important. Take the time and be honest.

STRONG COMMUNICATOR
Listen and Consider as Well as Speak

Please respond to each of the statements below based on how you currently "think" and how you "see the world" today by circling the most appropriate number (1-10) along the continuum.

Not at All   Sometimes   Always

1	2	3	4	5	6	7	8	9	10

1. When I communicate with others, I am aware of my inner monologue.

1	2	3	4	5	6	7	8	9	10

2. When I communicate with others, I make sure that I know my audience.

1	2	3	4	5	6	7	8	9	10

3. When I communicate with others, I am direct, specific and clear.

1	2	3	4	5	6	7	8	9	10

4. When I communicate with others, I pay attention to my nonverbal communication.

1	2	3	4	5	6	7	8	9	10

5. When I communicate with others, I listen more than I speak.

1	2	3	4	5	6	7	8	9	10

6. When I communicate with others, I am always positive and respectful.

Score: _____
(The sum of the numbers circled for each statement above)

STRONG COMMUNICATOR
Transformation Framework

1. When I communicate with others, I am aware of my inner monologue.	All good communication starts from a place of self-awareness. When Outstanding Leaders communicate with others, they are always aware of their inner monologue. They always assume the other person can read their mind and can sense if they are being discriminatory or appearing unconfident.
2. When I communicate with others, I make sure that I know my audience.	The best communication arises out of understanding your audience. Outstanding Leaders always know their audience's motivations, preferred communication styles, learning styles. This allows them to adapt their messages and increase the odds of effective communication.
3. When I communicate with others, I am direct, specific and clear.	Clear-cut communication increases the likelihood that people will comprehend and take action on whatever you are asking from them. Outstanding Leaders know that it is better to "over-explain" something than to leave room for misunderstanding.
4. When I communicate with others, I pay attention to my nonverbal communication.	Plenty of research suggests nonverbal communication is just as important as what a person says—maybe even more. Outstanding Leaders know that facial expressions, hand gestures, posture and eye contact all play a major role in affirming or undermining their messages.
5. When I communicate with others, I listen more than I speak.	One of the best ways to encourage open and honest communication within a group is to model active listening. Outstanding Leaders know that when someone is speaking to them, they should really listen to what is being said. They also ask follow-up questions to demonstrate that they are paying attention and to make sure there are no miscommunications. They also keep an open mind and focus on thoughtfully responding to what has been said. This practice builds rapport and understanding.

STRONG COMMUNICATOR
Transformation Framework

6. When I communicate with others, I am always positive and respectful.	Outstanding Leaders always prioritize transparent, fair and respectful communication within a group. They know that this is one of the best strategies for cultivating loyalty and boosting the team's performance. They do not hold their status over other people or use coercion or fear as motivators. Instead, they focus on bringing honest, positive and egoless attitudes to every situation that arises. Serving as a cheerleader instead of an autocrat helps maintain morale and can even facilitate creativity and effective problem solving.

"The art of communication is the language of leadership."

— **James Humes**

WHAT ARE YOUR THOUGHTS REGARDING YOUR CHALLENGES HERE?

Your thoughts on what you have just read are important. Take the time and be honest.

TRUSTWORTHY
Behaves in a Consistent, Highly Ethical and Fair Manner

Please respond to each of the statements below based on how you currently "think" and how you "see the world" today by circling the most appropriate number (1-10) along the continuum.

Not at All   Sometimes   Always

1	2	3	4	5	6	7	8	9	10

1. I always do what I say I am going to do.

1	2	3	4	5	6	7	8	9	10

2. I focus on having grace when I am under fire.

1	2	3	4	5	6	7	8	9	10

3. When needed, I always come to the rescue of others.

1	2	3	4	5	6	7	8	9	10

4. I tend to avoid gossip and cynicism.

1	2	3	4	5	6	7	8	9	10

5. I always set realistic targets for my teams and offer my help, if needed.

1	2	3	4	5	6	7	8	9	10

6. I find it easy to let go of control of a task or issue when warranted.

Score: _____
(The sum of the numbers circled for each statement above)

TRUSTWORTHY
Transformation Framework

1. I always do what I say I am going to do.	Nothing promotes trust more than keeping your word. Outstanding Leaders always uphold their end of the bargain. If circumstances change, as they often do, they communicate these changes to their team.
2. I focus on having grace when I am under fire.	No one respects a tyrant. Outstanding Leaders know that their team and colleagues are less likely to trust them if their primary mode of communication involves raising their voice, or humiliating others. When under pressure, Outstanding Leaders buy themselves some time by stepping back and taking a breath before saying something, they may regret.
3. When needed, I always come to the rescue of others.	When Outstanding Leaders sense that a team member or colleague is embarrassed by something he or she says in a meeting, they try to find a way to uphold any aspect of what the person is saying, and support it. In other words, they help the person save face.
4. I tend to avoid gossip and cynicism.	Outstanding Leaders know that their character becomes tarnished when they gossip. Sarcasm is another characteristic that can engender resentment and animosity. Outstanding Leaders know that people need to be listened to, appreciated, involved and connected. Transparent and direct communication tempered with respect, is key.
5. I always set realistic targets for my teams and offer my help, if needed.	Outstanding Leaders know that it is important to allow sufficient time for a project. They remain aware of the progress of a project and do not overload their team with unrealistic deadlines. If the task is too challenging or time-consuming, they ask how they can support the team.

TRUSTWORTHY
Transformation Framework

6. I find it easy to let go of control of a task or issue when warranted.	Outstanding Leaders know that when they let go of the need to control some things -- they open up the opportunities for their team members to take on more responsibility. They still hold the team members accountable and give productive feedback when necessary, but they allow the team to take the reins on occasion.
7. I routinely express gratitude to others.	Outstanding Leaders know that we all need to hear that our work is valued. They never underestimate the power of "thank you". They remember that what they appreciate becomes more appreciated.

"It takes two to do the trust tango--the one who risks (the trustor) and the one who is trustworthy (the trustee); each must play their role."

— Charles H. Green

WHAT ARE YOUR THOUGHTS REGARDING YOUR CHALLENGES HERE?

Your thoughts on what you have just read are important. Take the time and be honest.

CALM
Stays Cool Under Pressure, Especially When Things Go Wrong

Please respond to each of the statements below based on how you currently "think" and how you "see the world" <u>today</u> by circling the most appropriate number (1-10) along the continuum.

Not at All   Sometimes   Always

1	2	3	4	5	6	7	8	9	10

1. When facing challenging situations, I stop or pause first before responding.

1	2	3	4	5	6	7	8	9	10

2. In highly stressful situations, I automatically take a break.

1	2	3	4	5	6	7	8	9	10

3. I often calm down by taking time to reflect.

1	2	3	4	5	6	7	8	9	10

4. I normally refocus on the direction of all of my projects as progress is made.

1	2	3	4	5	6	7	8	9	10

5. I consider myself resolute and I hold high convictions in face of adversity.

1	2	3	4	5	6	7	8	9	10

6. I make it a point to speak to a mentor or trusted advisor when I face new and unfamiliar challenges.

1	2	3	4	5	6	7	8	9	10

7. When I find myself under stress, I try to change my self-talk.

Score: _____
(The sum of the numbers circled for each statement above)

CALM
Transformation Framework

1. When facing challenging situations, I stop or pause first before responding.	Outstanding Leaders will give themselves a little extra time before saying or doing something. It is very tempting in stressful situations to react quickly and say or do the first thing that comes to mind, however they know that it is more beneficial to pause first, then respond.
2. In highly stressful situations, I automatically take a break.	In highly stressful situations, Outstanding Leaders know that taking a break and allowing others involved to do the same, will dissipate any built up tension. It is quite amazing how much clearer they can think when they step away from something that has been bothering them.
3. I often calm down by taking time to reflect.	Outstanding Leaders know that self-reflection is a powerful skill for anyone to have. During reflection, it gives them the opportunity to check their attitude and determine whether what they are doing is what is required at that time. They, quite often, reflect to help them see things from different perspectives, which can help them reach solutions quicker.
4. I normally refocus on the direction of all of my projects as progress is made.	Outstanding Leaders know that the lack of clarity results in confusion, which can often cloud their judgement and their ability to make good decisions. Just asking, "What am I trying to accomplish right now?" helps them get back to what is most important at the time.
5. I consider myself resolute and I hold high convictions in face of adversity.	When challenged with a difficult issue requiring a prompt solution, Outstanding Leaders know that being caught up in one way of doing things can continue to create more pressure for them and those involved. When they start asking new questions, they will be forced to come up with new answers. In order to get new results, they have to take new actions and it starts with asking new questions.

CALM
Transformation Framework

6. I make it a point to speak to a mentor or trusted advisor when I face new and unfamiliar challenges.	Almost all the challenges we will ever face in life, someone has already faced. Outstanding Leaders know that instead of trying to work out everything on their own or even try to force something to happen, it can be beneficial to speak to a mentor or trusted advisor. They will be able to share new perspectives that will often help make the situation a lot easier to address.
7. When I find myself under stress, I try to change my self-talk.	Outstanding Leaders know that most stress is self-created and it often starts with their self-talk. If we consistently have negative self-talk, we will tend to experience negative emotions. Instead of being our own worst critic, it is more beneficial to change our self-talk to one that is encouraging, positive and focused on outcomes.

"Calm mind brings inner strength and self-confidence, so that's very important for good health."

— Dalai Lama

WHAT ARE YOUR THOUGHTS REGARDING YOUR CHALLENGES HERE?

Your thoughts on what you have just read are important. Take the time and be honest.

OPEN-MINDED
Invites Constructive Dissent or Disagreements

Please respond to each of the statements below based on how you currently "think" and how you "see the world" <u>today</u> by circling the most appropriate number (1-10) along the continuum.

Not at All   Sometimes   Always

1	2	3	4	5	6	7	8	9	10

1. I am open to accepting as well as giving.

1	2	3	4	5	6	7	8	9	10

2. I am willing to embrace change.

1	2	3	4	5	6	7	8	9	10

3. I have a high level of curiosity.

1	2	3	4	5	6	7	8	9	10

4. I am willing to try new things.

1	2	3	4	5	6	7	8	9	10

5. I do not judge people or ideas without first listening attentively.

1	2	3	4	5	6	7	8	9	10

6. I am always open to other's opinions.

1	2	3	4	5	6	7	8	9	10

7. I always respect differences in people and personalities.

1	2	3	4	5	6	7	8	9	10

8. I primarily live in the present and not the past.

1	2	3	4	5	6	7	8	9	10

9. I routinely turn problems into opportunities.

Score: _____
(The sum of the numbers circled for each statement above)

OPEN-MINDED
Transformation Framework

1. I am open to accepting as well as giving.	Outstanding Leaders have the ability to think of things from various angles and viewpoints. This allows them to accept more and take advantage of different ways to accomplish needed outcomes.
2. I am willing to embrace change.	All Outstanding Leaders are willing and able to embrace change with open arms because they see various possibilities and outcomes, rather than judging it from only one angle of "experience".
3. I have a high level of curiosity.	All Outstanding Leaders have a natural ability to want to know more about something. They tend not to judge an idea that may be presented to them with only what they know. They prefer to inquire, learn and discover new insights.
4. I am willing to try new things.	Outstanding Leaders value new ideas and trying new things. They do not mind trying out a new place for dinner, or making new friends, or embracing a new project.
5. I do not judge people or ideas without first listening attentively.	Outstanding Leaders are willing to listen to someone without judging them or jumping to conclusions before they have finished.
6. I am always open to other's opinions.	It is okay to be opinionated. However, Outstanding Leaders have a good understanding of the fact that they do not have to agree with everyone or vice versa.
7. I always respect differences in people and personalities.	All Outstanding Leaders are open to people's values, beliefs, and differences. They believe that differences are what make everyone unique—which is a valuable quality.
8. I primarily live in the present and not the past.	Outstanding Leaders do not dwell on the past. They trust that everything happens for a reason. Furthermore, they try not to fret over the future because they see so many possible outcomes. Focusing on the present and appreciating those moments is seen as far more sensible to them.

OPEN-MINDED
Transformation Framework

9. I routinely turn problems into opportunities.	Outstanding Leaders acknowledge the fact that every problem has a solution or at least a different perspective that is better than the first. This enables them to see clearly when faced with difficult situations and tackle problems without panicking. For them, problems are opportunities to learn what to fix.

"Without an open-minded mind, you can never be a great success."

— Martha Stewart

WHAT ARE YOUR THOUGHTS REGARDING YOUR CHALLENGES HERE?

Your thoughts on what you have just read are important. Take the time and be honest.

PROFESSIONAL
Are Respectable both at Work and in Private Life

Please respond to each of the statements below based on how you currently "think" and how you "see the world" <u>today</u> by circling the most appropriate number (1-10) along the continuum.

Not at All　　　　　　　　　　　　　　Sometimes　　　　　　　　　　　　　　Always

1	2	3	4	5	6	7	8	9	10

1. I make sure that my appearance is always neat and appropriate.

1	2	3	4	5	6	7	8	9	10

2. I always maintain a strong, yet reassuring, demeanor.

1	2	3	4	5	6	7	8	9	10

3. I consider myself a very reliable person.

1	2	3	4	5	6	7	8	9	10

4. I maintain a high level of competence in all areas of my profession.

1	2	3	4	5	6	7	8	9	10

5. I make sure that all of my actions are ethical.

1	2	3	4	5	6	7	8	9	10

6. I always maintain my poise, even in challenging situations.

1	2	3	4	5	6	7	8	9	10

7. I am always cognizant of my phone, email and texting etiquette.

1	2	3	4	5	6	7	8	9	10

8. I make sure that all of my written correspondence is concise and representative of my standards.

1	2	3	4	5	6	7	8	9	10

9. I take pride in my organizational skills.

1	2	3	4	5	6	7	8	9	10

10. I always hold myself accountable for my actions.

Score: _____
(The sum of the numbers circled for each statement above)

PROFESSIONAL
Transformation Framework

1. I make sure that my appearance is always neat and appropriate.	Outstanding Leaders are always neat in appearance. They make sure to meet or even exceed the requirements of their organization's dress code, and pay special attention to their appearance when meeting with stakeholders and clients.
2. I always maintain a strong, yet reassuring, demeanor.	An Outstanding Leader's demeanor always exudes confidence but not cockiness. They are polite and well-spoken whether they are interacting with customers, superiors or co-workers. They keep their calm, even during tense situations.
3. I consider myself a very reliable person.	Outstanding Leaders can be counted on to find a way to get the job done. They respond promptly and follow through on promises in a timely manner.
4. I maintain a high level of competence in all areas of my profession.	Outstanding Leaders strive to become experts in their field, which sets them apart from others. They maintain their focus as life-long learners by taking courses, attending seminars and attaining any related professional designations.
5. I make sure that all of my actions are ethical.	Outstanding Leaders go out of their way to adhere to a strict code of ethics and display ethical behavior at all times.
6. I always maintain my poise, even in challenging situations.	Outstanding Leaders maintain their poise even when facing a difficult situation.
7. I am always cognizant of my phone, email and texting etiquette.	Outstanding Leaders know that their etiquette is an important component of professional behavior. They also try not to dominate the conversation and listen intently to the other party.
8. I make sure that all of my written correspondence is concise and representative of my standards.	Outstanding Leaders keep their emails, text messages and letters brief and to the point. Their tone is always polite and formal without being "stuffy".

PROFESSIONAL
Transformation Framework

9. I take pride in my organizational skills.	Outstanding Leaders can quickly and easily find what is needed. Their work area is generally neat and organized.
10. I always hold myself accountable for my actions.	Outstanding Leaders are accountable for their actions at all times. If they make a mistake, they own up to it and try to fix it, if possible. They do not try to place the blame on a colleague. If their organization makes the mistake, they take responsibility and work to resolve the issue.

"A professional is someone who can do his best work when he doesn't feel like it."

— Alistair Cooke

WHAT ARE YOUR THOUGHTS REGARDING YOUR CHALLENGES HERE?

Your thoughts on what you have just read are important. Take the time and be honest.

CLEAR THINKING
Simplifies Things to Maximize Understanding and Minimize Confusion

Please respond to each of the statements below based on how you currently "think" and how you "see the world" <u>today</u> by circling the most appropriate number (1-10) along the continuum.

Not at All				Sometimes					Always

1	2	3	4	5	6	7	8	9	10

1. I often translate complex messages into a language that can be more easily understood.

1	2	3	4	5	6	7	8	9	10

2. I routinely break long messages and directions into smaller parts.

1	2	3	4	5	6	7	8	9	10

3. When a problem has portions that are complex, I make a list and review each individually.

1	2	3	4	5	6	7	8	9	10

4. I will engage an expert to help communicate subject matter that requires a deeper understanding.

1	2	3	4	5	6	7	8	9	10

5. I always remind myself to focus on the big picture.

1	2	3	4	5	6	7	8	9	10

6. I will often create a map when I am leading my team through problem solving.

1	2	3	4	5	6	7	8	9	10

7. With complex projects, I take the lead in helping my team digest the challenge "in small bits".

Score: _____
(The sum of the numbers circled for each statement above)

CLEAR THINKING
Transformation Framework

1. I often translate complex messages into a language that can be more easily understood.	Outstanding Leaders know that languages are constantly changing and younger generations as well as experts add new words often. Trying to understand complex theories or technical jargon can be a huge pain. Outstanding Leaders take a minute or two to think about what is the underlying theme or message and translate the language into something that any adolescent can understand.
2. I routinely break long messages and directions into smaller parts.	As they grow, projects, problems, and challenges can become so complex that it is difficult to even know where to start when trying to accomplish something. Outstanding Leaders tend to simplify things by breaking them down into small parts so that team members can look at each facet, one at a time. Once divided, they can prioritize and create plans more efficiently.
3. When a problem has portions that are complex, I make a list and review each individually.	When parts of a problem are beyond an Outstanding Leader's direct knowledge or skillset, he or she will identify the crucial issues and attempt to understand them or solve the problem. He or she will present the issues to the entire team to benefit from the shared expertise.
4. I will engage an expert to help communicate subject matter that requires a deeper understanding.	When Outstanding Leaders do not understand how something works, they will get someone who does. By bringing in an expert for support, Outstanding Leaders can then focus their efforts where they are strong and speed their process.
5. I always remind myself to focus on the big picture.	When Outstanding Leaders have a complicated issue, they do not get caught up in the details. They take a moment to step back and look at the big picture. The larger perspective may reveal that the small problems are merely symptoms and can be easily solved with a bigger, simpler systemic solution.

CLEAR THINKING
Transformation Framework

6. I will often create a map when I am leading my team through problem solving.	Visuals can be very useful in problem solving and Project Leadership. Outstanding Leaders will create a visual path to help their team understand the plan. The map makes it easier for them and team members to track progress and apply the appropriate energy.
7. With complex projects, I take the lead in helping my team digest the challenge "in small bits".	When you are starting at a huge or complex project, you may be tempted to attack it on all fronts or you may not even know where to start. The only way to "eat an elephant" is one bite at a time. Outstanding Leaders allow his or her team to start with a small step and to learn as they go. It will seem slow and nearly insurmountable at first. However, before long, they gain the understanding and proficiency that will carry them through to the end with increased momentum.

"Clear thinking requires courage rather than intelligence."

— Thomas Szasz

WHAT ARE YOUR THOUGHTS REGARDING YOUR CHALLENGES HERE?

Your thoughts on what you have just read are important. Take the time and be honest.

NURTURING
Shows Respect for All People, their Opinions, and their Capacity to Achieve

Please respond to each of the statements below based on how you currently "think" and how you "see the world" today by circling the most appropriate number (1-10) along the continuum.

Not at All   Sometimes   Always

1	2	3	4	5	6	7	8	9	10

1. I am supportive of team members seeking more professional training and development.

1	2	3	4	5	6	7	8	9	10

2. I frequently look for opportunities to provide team members increased responsibility.

1	2	3	4	5	6	7	8	9	10

3. I always "connect the dots" regarding how each team member brings value to the project.

1	2	3	4	5	6	7	8	9	10

4. I always recognize achievements and show my appreciation for good performance.

1	2	3	4	5	6	7	8	9	10

5. I find ways to create a learning project culture.

1	2	3	4	5	6	7	8	9	10

6. I always establish and clearly communicate project and team goals.

1	2	3	4	5	6	7	8	9	10

7. I promote professional growth among my team members by encouraging and providing stretch goals.

1	2	3	4	5	6	7	8	9	10

8. I consider myself a role model to my team members.

1	2	3	4	5	6	7	8	9	10

9. I make sure that I do not neglect poor performance.

Score: _____
(The sum of the numbers circled for each statement above)

NURTURING
Transformation Framework

1. I am supportive of team members seeking more professional training and development.	Outstanding Leaders know that each team member brings some level of experience and knowledge to the project. They also know that effective training is how to best integrate a team member's existing knowledge into the project's plan and strategy.
2. I frequently look for opportunities to provide team members increased responsibility.	Outstanding Leaders know that providing team members with opportunities to take on additional responsibility will show them how much they trust them.
3. I always "connect the dots" regarding how each team member brings value to the project.	Outstanding Leaders make it a point to help "connect the dots" regarding each team member's contribution to the project. Connecting the dots means explaining why each team member and task is extremely important to the mission of the project.
4. I always recognize achievements and show my appreciation for good performance.	Outstanding Leaders are committed and dedicated to the project team. Thanking their team and giving honest recognition for the team's work achievements can help the team to feel appreciated and enhance their job satisfaction.
5. I find ways to create a learning project culture.	Professional development and growth happens where people are encouraged to be curious and learn. Outstanding Leaders provide feedback regularly and allow team members the space to explore new ideas or solutions. Not only does a learning-oriented culture help to cultivate leadership, it also breeds innovation and high-performance.
6. I always establish and clearly communicate project and team goals.	How can you prepare your team members for leadership if you don't know what their goals are? In addition to giving them a better understanding of how each individual fits into the Team as a whole, Outstanding Leaders also take time to understand personal goals.

NURTURING
Transformation Framework

7. I promote professional growth among my team members by encouraging and providing stretch goals.	Professional growth is uncomfortable and challenging. Outstanding Leaders know that creativity and innovation are born out of the need to stretch beyond a team member's comfort zone to accomplish a goal and that achieving stretch-goals successfully can be a great confidence booster.
8. I consider myself a role model to my team members.	Intentionally or not, the Leaders within an organization demonstrate the characteristics they cultivate in their teams. Outstanding Leaders know that it is important for them to intentionally model the kinds of behavior they expect from all team members.
9. I make sure that I do not neglect poor performance.	Poor performance is usually a sign of disengagement and that something is not working. Outstanding Leaders know that it is important to provide the team members with the opportunity to improve, along with actions to be taken. This does not have to be a punitive experience. If Outstanding Leaders see everyone with the potential of improving their performance, they help everyone to do their best work.

"Management is about arranging and telling. Leadership is about nurturing and enhancing."

— Tom Peters

WHAT ARE YOUR THOUGHTS REGARDING YOUR CHALLENGES HERE?

Your thoughts on what you have just read are important. Take the time and be honest.

SUPPORTIVE
Helps Others Build on Their Strengths and Shows Appreciation

Please respond to each of the statements below based on how you currently "think" and how you "see the world" today by circling the most appropriate number (1-10) along the continuum.

Not at All · Sometimes · Always

1	2	3	4	5	6	7	8	9	10

1. I am supportive and encourage team members to support each other.

1	2	3	4	5	6	7	8	9	10

2. I always expect and encourage teamwork.

1	2	3	4	5	6	7	8	9	10

3. I am committed to my team's success and help grow the commitment of each team member.

1	2	3	4	5	6	7	8	9	10

4. I always create a team vision and help people personalize it.

1	2	3	4	5	6	7	8	9	10

5. I always focus on relationships and encourage others to do the same.

1	2	3	4	5	6	7	8	9	10

6. I make myself available to help and let my team grow independently, when possible.

Score: _____
(The sum of the numbers circled on each statement above)

SUPPORTIVE
Transformation Framework

1. I am supportive and encourage team members to support each other.	Outstanding Leaders are supportive of both the team as a whole as well as the individuals on the team. They never forget that a team is made up of individuals and when they support them, they build team confidence and create positive attitudes. Since they know that confidence and a positive attitude will improve individual (and team) results, they not only do this, but also help people do the same for each other.
2. I always expect and encourage teamwork.	It is difficult to expect people to come together as an effective team if there is not a clear and definitive expectation of the importance. Outstanding Leaders know that building great teams start by making expectations clear. Outstanding Leaders make sure they are encouraging teamwork through their conversations, feedback, recognition, reward systems and more.
3. I am committed to my team's success and help grow the commitment of each team member.	Outstanding Leaders know that the best teams are committed to their success and to each other. Outstanding Leaders recognize the importance of this commitment and engagement. They know that their involvement may require conversations, coaching and even conflict resolution. They know that helping their teams become more committed to the work and each other will pay huge dividends in results.
4. I always create a team vision and help people personalize it.	Outstanding Leaders know that a team will be committed and do great work, when they move in the same direction for overall project success. Outstanding Leaders set achievable goals and involve team members in the goal setting process. They help people connect their individual tasks to the goals of the team and the project vision.

SUPPORTIVE
Transformation Framework

5. I always focus on relationships and encourage others to do the same.	Outstanding Leaders know that relationships among team members matter and will aid in team development as well as success. They know that to maintain highly successful teams, you must allow time and space for team members to build relationships while they accomplish tasks.
6. I make myself available to help and let my team grow independently, when possible.	Outstanding Leaders know that they must make time for and invest time in their team. However, they do not micro-manage. They do help team members grow and learn by giving them space and opportunity.

"None of us is as smart as all of us."

— Ken Blanchard

WHAT ARE YOUR THOUGHTS REGARDING YOUR CHALLENGES HERE?

Your thoughts on what you have just read are important. Take your time and be honest.

DETERMINE YOUR LEADERSHIP ATTRIBUTE GAPS

First, you should complete the table below by including your scores from each of the 10 exercises and calculating "Your Gap" & Gap Percentage (%).

Project Leadership Attribute	Max Score (A)	Your Score (B)	Your Gap (C) = [A-B]	Your Gap % (C)/(A) Gap % = 0, If (C) = 0
Example	100	70	30	30/100 = 30%
1. Leads by Example	100			
2. Visionary	100			
3. Strong Communicator	60			
4. Trustworthy	60			
5. Calm	70			
6. Open-Minded	90			
7. Professional	100			
8. Clear Thinking	70			
9. Nurturing	90			
10. Supportive	60			

Now, complete the exercises on the following pages.

Part A will assist in identifying your "Lagging" and "Leading" Attributes

Part B will assist in developing a plan to improve the "Lagging" attributes during the transformation process.

Part C will assist in constructing your Personal Leadership Vision.

IDENTIFY YOUR "LAGGING" AND "LEADING" ATTRIBUTES

Part A - Identify your Lagging and Leading Attributes

Complete the table below by filling in all three columns as indicated.

Project Leadership Attribute	Your Gap Percentages (From Page 57)	Your Lagging Attributes (X) Enter (X) if Gap % is equal to or <u>greater than</u> 50%	Your Leading Attributes (X) Enter (X) if Gap % is <u>Less than</u> 50%
1. Leads by Example			
2. Visionary			
3. Strong Communicator			
4. Trustworthy			
5. Calm			
6. Open-Minded			
7. Professional			
8. Clear Thinking			
9. Nurturing			
10. Supportive			

Your *Lagging Attributes* present opportunities for improving your ability to think and perform as an Outstanding Leader. Significantly demonstrating improvement in your ability to enhance these attributes in your leadership style and focus will be required for a successful transformation.

During the initial stages of your transformation, you should rely heavily on using your *Leading Attributes (The attributes that it appears you are currently effectively deploying)*. This will help *compensate* for the Leadership attributes that require strengthening.

YOUR THOUGHTS REGARDING YOUR LEADING AND LAGGING ATTRIBUTES?

Take your time and be honest.

> *"All growth depends upon activity. There is no development physically or intellectually without effort, and effort means work."*
>
> *— Calvin Coolidge*

DEVELOP YOUR LAGGING ATTRIBUTE IMPROVEMENT PLAN

Part B - Develop your plan to improve the Lagging Attributes during the transformation process.

Take as much time as needed to thoughtfully and realistically complete the table below. By demonstrating documented improvement in effectively deploying these Attributes, you will significantly enhance your ability to successfully complete your transformation. Your overall Transformation Action Plan execution schedule should include the time you believe it will take for you to improve the effective and consistent use of these Attributes.

List Your Lagging Attributes Here	Attributes To Be Improved (X)	How You Will Improve	Time Required (Months)
1.		▪ ▪ ▪	
2.		▪ ▪ ▪	
3.		▪ ▪ ▪	

DEVELOP YOUR LAGGING ATTRIBUTE IMPROVEMENT PLAN

(Continued)

List Your Lagging Attributes Here	Attributes To Be Improved (X)	How You Will Improve	Time Required (Months)
4.		<ul><li></li><li></li><li></li></ul>	
5.		<ul><li></li><li></li><li></li></ul>	
6.		<ul><li></li><li></li><li></li></ul>	
7.		<ul><li></li><li></li><li></li></ul>	

CONSTRUCT YOUR PERSONAL LEADERSHIP VISION

Part C - Constructing Your Personal Leadership Vision

A Personal Leadership Vision is one that focuses action, provides direction and inspires you and your stakeholders. It also encompasses personal "affirmations" to provide daily focus, encouragement and inner support throughout your transformation journey.

Sharing this vision with others is a powerful way to unearth your real passions and increase your energy and capacity to successfully make the transformation from good *Project Manager* to *Outstanding Project Leader*.

Personalize the vision statement below by inserting the Lagging and Leading Attributes you have identified.

"I will live each day as though I have all the power and influence necessary to become an Outstanding Project Leader to the benefit of myself, my project teams and my organization. Through listening to and valuing others, I will consistently think like a Leader and gain a Leader's perspective. I will strive to gain mastery over Project Leadership through increasing my Leadership skills, expanding my circle of influence and increasing my capacity to take advantage of lessons-learned shared by my mentors, contemporary literature and others with demonstrated Leadership expertise.

I will utilize my strongest Leadership Attributes which include: (Your Leading Attributes Here)_____, _____, _____, _____, _____, _____ and I will focus on improving my ability to *(Your Lagging Attributes Here)* _____, _____, _____, _____ and _____.

I will improve by being open-minded and actively seeking all available sources for learning and growth in these areas.

I choose to focus on being both a highly effective Leader and a respected contributor to my organization."

LIST YOUR TRANSFORMATION AFFIRMATIONS

Your Personal Leadership Vision is a great place to reference when constructing a set of personal and professional "affirmations". The affirmations can be used to provide daily focus, encouragement and inner support throughout your transformation journey.

Review your Personal Leadership Vision and consider all of the thoughts and ideas you have had to this point regarding how to successfully complete your transformation. Then, list at least four affirmations at this time. You can always add to this list later.

(Example: I have the strength and courage to share my vision with anyone.)

1.
2.
3.
4.
5.
6.
7.

Analyzing Your Current Organizational Alignment, Project Leadership Tendencies and Development Perspective

The first step in determining where you want to be is to understand where you are today. In the words of the baseball great Yogi Berra, *"If you don't know where you're going, you might not get there"*. Knowing what you are doing or not doing today allows you to identify both strengths and weaknesses. The analysis should involve using a performance-tracking tool and keeping a record or diary to reflect an accurate picture of your reality.

Let us start with you completing the following **Project Leadership Assessment**. This set of exercises is designed to identify and explore your:

- Current Organizational Alignment;
- Leadership Type;
- Leadership Behaviors;
- Leadership Resilience; and
- Leadership Development Perspective.
  (i.e. what is possible based on known strengths and deficiencies).

This insight will aid in the development of your Transformation Action Plan and activities required to improve your overall Leadership skills and capacity.

Make sure that you take the time to completely respond to the *"What are your thoughts at this time?"* question at the end of each exercise. This important feedback will be used later.

Now, let's move on to the Project Leadership Assessment.

PROJECT LEADERSHIP ASSESSMENT

The following five self-assessment exercises are designed to help you identify your own Project Leadership scores. The assessments target five areas of Leadership. As a composite, they will give you a general sense of where to focus your efforts to improve your Project Leadership capacity. As you progress through the workbook and coaching sessions, you will gain more insight into your assessment results. This valuable insight will provide a more in-depth and thorough analysis of your current capacity and potential for Project Leadership growth.

We encourage you to consider these assessments as a way to get a snapshot of where you excel and where you may want to focus your developmental activities and energies.

SELF-ASSESSMENT EXERCISES
What Is Your Current Organizational Alignment?
What Is Your Leadership Type?
What Are Your Leadership Behaviors?
What Is Your Leadership Resiliency?
What Is Your Leadership Development Perspective?

WHAT IS YOUR CURRENT ORGANIZATIONAL ALIGNMENT?

Score Yourself on Managing Alignment of Self and Organization

Think about your level of response to work situations during the past six months:

| Never (1) | Rarely (2) | Sometimes (3) | Often (4) | Almost always (5) |

1.	I am aware of my own passions and values.	1	2	3	4	5
2.	My behavior consistently reflects my goals and values.	1	2	3	4	5
3.	I feel safe pushing back when I am asked to do things that are not aligned with my values.	1	2	3	4	5
4.	I am aware that my behavior and decisions have a significant impact on the organization's performance and culture.	1	2	3	4	5
5.	I am deliberate about aligning my organization's pay and performance systems with the types of behaviors I want to encourage (both results and behaviors).	1	2	3	4	5
6.	My organization's key measures and systems encourage the right actions aligned with the culture, and discourage actions that will impact my performance or make me uncomfortable.	1	2	3	4	5
7.	I am aware of how my values align with those of my organization and where there are misalignments.	1	2	3	4	5
8.	I take steps to encourage changes in my organization's culture such as talking about our values and reinforcing what we say we care about.	1	2	3	4	5

Your Current Organizational Alignment

Your Total Score _____

(The sum of the numbers circled on each statement on the previous page)

If your overall score in this exercise is 24 or less, it is time to pay attention to your alignment with your organization and also the alignment of culture and systems within the organization that you are able to impact.

If your overall score in this exercise is 25–31, you are in the healthy range, but could still benefit from some focus on alignment.

If your score is 32 or above, Congratulations! You are well aligned with your organization and organization's culture and systems are well aligned.

What are your thoughts regarding your organizational alignment at this time?

WHAT IS YOUR LEADERSHIP TYPE?

Score Yourself on Awareness of Leader Type and Self-Management

Think about your level of response to work situations during the past six months:

| Never (1) | Rarely (2) | Sometimes (3) | Often (4) | Almost always (5) |

#						
1.	I have taken a leadership type assessment such as the Enneagram, Myers-Briggs Type Indicator or DISC and used this information about myself to increase my effectiveness.	1	2	3	4	5
2.	I understand my gifts and limitations, and try to leverage my strengths and manage my limitations.	1	2	3	4	5
3.	I have a reflection practice where I understand, actively monitor and work with my "fixations" (a fixation is a negative thought pattern).	1	2	3	4	5
4.	I have a clear sense of who I am and what I want to contribute to my organization and in the world.	1	2	3	4	5
5.	I manage my emotional reactions to allow me to respond with socially appropriate behavior.	1	2	3	4	5
6.	I am aware of what causes me stress and actively manage it.	1	2	3	4	5
7.	I have positive coping strategies.	1	2	3	4	5
8.	I actively seek ways to feel empowered even when my organization may not give me this sense of empowerment.	1	2	3	4	5

Leader Type and Self-Management

Your Total Score _____

(The sum of the numbers circled on each statement on the previous page)

If your overall score in this exercise is 24 or less, it is time to pay attention to your leadership type and self-management.

If your overall score in this exercise is 25–31, you are in the healthy range, but could still benefit from some focus on your leadership type and self-management.

If your overall score is 32 or above, Congratulations! You are self-aware and using your leadership type to increase your effectiveness.

What are your thoughts regarding your leadership type at this time?

WHAT ARE YOUR LEADERSHIP BEHAVIORS?

Score Yourself on Leadership Behaviors

Think about your level of response to work situations during the past six months:

| Never (1) | Rarely (2) | Sometimes (3) | Often (4) | Almost always (5) |

#							
1.	I tend to be proactive—I anticipate what is coming next and actively manage it. (Depending on your current role, this may happen primarily at this time in your personal life.)	1	2	3	4	5
2.	I focus on creating results in a way that encourages others to grow and develop while accomplishing their tasks.	1	2	3	4	5
3.	I think about the impact of my actions on the overall organization rather than just getting the job done.	1	2	3	4	5
4.	I see how my work contributes to the overall organizational success and deliberately try to improve my organization and myself.	1	2	3	4	5
5.	I take time to mentor others—even when I am busy.	1	2	3	4	5
6.	I consider myself a life-long learner because of the time I spend reading and trying new ideas and activities.	1	2	3	4	5
7.	I have the courage to speak out in a professional manner when asked to do something I disagree with.	1	2	3	4	5
8.	I accomplish results by working with and through others in a positive and constructive manner.	1	2	3	4	5

Leadership Behaviors

Your Total Score _____

(The sum of the numbers circled on each statement on the previous page)

If your overall score in this exercise is 24 or less, it is time to pay attention to your leadership behaviors and look for ways to develop in alignment with your goals.

If your overall score in this exercise is 25–31, you are in the healthy range, but could still benefit from some focus on your leadership behaviors.

If your score is 32 or above, Congratulations! You are likely performing well in the area of leadership behaviors.

What are your thoughts regarding your Leadership behaviors at this time?

WHAT IS YOUR LEADERSHIP RESILIENCE?

Score Yourself on Leadership Resilience

Think about your level of response to work situations during the past six months:

Never (1)	Rarely (2)	Sometimes (3)	Often (4)	Almost always (5)

1.	I consistently take care of my physical needs such as getting enough sleep and exercise.	1	2	3	4	5
2.	I have a sense of purpose and get to do activities that contribute to that purpose daily.	1	2	3	4	5
3.	I have a high degree of self-awareness and actively manage my thoughts.	1	2	3	4	5
4.	I have a strong support system consisting of a healthy mix of friends, colleagues and family.	1	2	3	4	5
5.	I can reframe challenges to find something of value in most situations.	1	2	3	4	5
6.	I build strong trusting relationships at work.	1	2	3	4	5
7.	I am aware of my own self-talk and actively manage it.	1	2	3	4	5
8.	I have a professional development plan that includes gaining skills and acquiring additional perspectives.	1	2	3	4	5

Leadership Resilience

Your Total Score _____

(The sum of the numbers circled on each statement on the previous page)

If your overall score in this exercise is 24 or less, it is time to pay attention to your resilience.

If your overall score in this exercise is 25–31, you are in the healthy range, but could still benefit from some focus on resilience.

If your score is 32 or above, Congratulations! You are likely performing well in the area of resilience.

What are your thoughts regarding your Leadership resilience at this time?

WHAT IS YOUR LEADERSHIP DEVELOPMENT PERSPECTIVE?

Score Yourself on Developmental Perspective Aligned with Project Leadership

Think about your level of response to work situations during the past six months:

| Never (1) | Rarely (2) | Sometimes (3) | Often (4) | Almost always (5) |

#		1	2	3	4	5
1.	I have a sense of life purpose and do work that is generally aligned with that purpose.	1	2	3	4	5
2.	I am motivated by the impact I make on the world more than on personal notoriety.	1	2	3	4	5
3.	I try to live my life according to my personal values.	1	2	3	4	5
4.	I believe that collaboration across groups and organizations is important to accomplish our goals.	1	2	3	4	5
5.	I believe that getting business results must be balanced with treating people fairly and kindly as well as have an impact on our customers and community.	1	2	3	4	5
6.	I seek input from others consistently to test my thinking and expand my perspective.	1	2	3	4	5
7.	I think about the impact of my work on the many elements of our community and beyond.	1	2	3	4	5
8.	I am open-minded and curious, always trying new things and learning from all of them.	1	2	3	4	5
9.	I appreciate the value of rules. Yet, I am willing to question them in a professional manner in order to meet project goals and improve the service we provide to our stakeholders.	1	2	3	4	5

Your Alignment with Project Leadership

Your Total Score_____

(The sum of the numbers circled on each statement on the previous page)

If your overall score in this exercise is 27 or less, it's time to pay attention to your developmental level and to focus on developing in the area of developmental perspective (*a sense of what advancement is possible based on your known strengths and deficiencies*).

If your overall score in this exercise is 28–35, you are in the healthy range, but could still benefit from some focus on developing in the area of developmental perspective.

If your score is 36 or above, Congratulations! Your developmental level appears to be aligned with Project leadership, yet this assessment is only a subset of a full assessment.

What are your thoughts regarding your alignment with Project Leadership at this time?

3.

Planning Your Transformation Journey

Many people yearn for personal or professional transformation without knowing how to start such a challenging change in their lives. You can easily waste your time and energy on false starts or take a few steps in the right direction, only to find that old habits and past conditioning pull you back to where you began. Your successful journey from "Project Manager" to "Project Leader" requires a thoughtful and realistic transformation plan.

Natural instincts and leadership talent alone, no matter how robust, cannot instantly obtain the transformation from "Manager" to "Leader". There is a proven science to executing this type of transformation and it requires a well thought out plan.

However, with a good plan, transformations of this type are still extremely challenging for two distinct reasons. First, the actual "future state" is unknown when you begin, and is determined through some trial and error as new information and resources are gathered. This makes it impossible to "manage" a transformation with static or linear plans. You can have an over-arching transition strategy, but the actual transformation process literally must "emerge" as you go. Secondly, if the future state is so radically different from the current state, you must spend more time to really understand yourself, your organization and the culture within which you operate prior to attempting the execution of even the best plan.

This section is designed to help you envision and articulate how you will approach each of the ten areas associated with the action planning process in Part IV of this program.

By thinking through and documenting the approach that "best" fits within your current organizational situation, your current relationships and the time commitment you are willing to make will ensure that a realistic action plan, implementation schedule and tracking process can be developed.

Let's get serious and start planning your journey.

1. How would you briefly describe what you are about to do, why are you doing it and what benefits do you wish to gain?

2. How will you create a sense of urgency within yourself and maintain it throughout your transformation process?

3. How will you approach putting your Transformation Support Team in place?

4. How will you engage your key stakeholders - particularly managers, family and friends - in helping you make this transformation happen?

5. How will you approach "over-communicating" your Personal Leadership Vision?

6. How will you approach the removal of obstacles to your transformation success?

7. Are you comfortable with cultivating new ideas, learning from others and changing personal and professional behaviors?

    If you are comfortable..."Why?"   If you are not..."Why not?"

8. How will you go about realigning relationships (if required) to enable your vision and support your transformation?

9. Why do you believe that you can upgrade your leadership skills to the extent required to successfully complete your transformation?

Imagining and Constructing Your Transformation Support Team and Communications Strategy

It is impossible to achieve a major self-development goal without managing and getting the right level of support from all those around you. Success is not a solo sport. It takes a team of people supporting you along the way. If you take a conscious approach to seeking out the support you will need, rather than leaving it to chance, your path will be smoother and your success more likely. This is true of all successes…including the transformation from "Project Manager" to "Project Leader".

The success or failure of a transformation of this type often rests on how well your support team understands what you are asking them to do and how well you communicate your personal vision and support needs. This will determine how well each member of your support team conveys your vision and needs to others.

Therefore, it is important to take the time to imagine everyone you will engage during your transformation period, construct a transformation support team and develop an effective communications strategy. However, you should remember that not everyone you know and engage will or should be on your support team.

With this in mind, in this section we will share with you six unique character types that we have found to play key roles in supporting both personal and professional transformations. If carefully selected, each of these types of individuals will provide you with important, objective, factual and emotional support.

Following the discussion of each character type, you will be asked to take an initial cut at listing potential team members who fit the character type. You should also thoughtfully include the relevant information requested in the exercise regarding each potential member. This insight will be used in Part IV of this workbook.

You will be asked later in this section to construct an outline of a communications strategy that could be used to ensure accurate and effective communications among those who can either "promote" or "derail" your transformation efforts.

Remember, you will have the opportunity to re-visit and finalize your potential Support Team Member list and the communications strategy in Part IV of this workbook. Therefore, you should focus here on developing an exhaustive list of potential members for each character type as well as a succinct communications strategy outline.

Okay. Let's get to work.

Imagining and Constructing
Your Transformation Support Team

Our sense of imagination makes it possible for us to experience a whole world inside our minds. It gives us the ability to look at any situation from a different point of view and to mentally explore the past and the future.

Just imagine that you are asked to coach an all-star baseball team and were given the opportunity to select the greatest players of all time to fill each of the nine positions.

How would you approach selecting your team? Chances are you would first take the time to understand the role that each of the nine positions played on the team and the type of personality that would best fit that role. This ensures that you would have all of the bases covered with the optimum type of individual and personality.

However, to maximize your success in actually winning the game, you would also want to find out if the individual is willing to become a part of your team, has the time to be an effective team member and whether or not he or she would get along with the rest of the team.

This same approach is how you should imagine who you should ask to join your Transformation Support Team and the specific role you will ask each member to play. Your Support Team must be more than a group of individuals. To be effective, your Team must be able to work collectively toward one common goal --- supporting your transformation from "Manager" to "Leader".

Each of your Support Team members must know the role that you wish them to play. By accepting the role you define, each team member will tend to behave, contribute and interact with each other, and with stakeholders, in a consistent and productive fashion.

However, it is important and up to you to make sure that they clearly understand what is needed to ensure your success and that they feel that they are in the position, are willing to be on your team, and have the time to provide what is needed.

In the following exercise, you will be provided a description of each of the six character types that we have found can play a key role in any professional transformation.

Then, you will be asked to make a list of the people you know (inside or outside your current workplace) who might be the best potential candidates --- based on the character description, expertise and level of support that you will need from them.

You will also be asked to "rank" them based of your preference of selection, given the choice.

Remember, you will have the opportunity to re-visit and finalize your potential Support Team Member list in Part IV of this workbook. So, at this point include everyone and anyone who you feel matches the character types and would possibly say "yes" to become a part of your Support Team.

"Great things in business are never done by one person. They're done by a team of people."

— Steve Jobs

The Visionary: Everybody needs someone in their life who cannot help but aim for the stars.

The Visionary has good communication skills. He or she can help you verbalize your personal vision and help you explain your needs to the rest of the Support Team. In addition, the visionary, by nature, is an active listener and can help you incorporate your Leadership development needs and provide insights into how to involve others in reaching your transformation milestones. Since most visionaries are strategic planners, he or she can also help in validating your action plan and communications strategy.

Name /Position/Title	Current Relationship?	How would you rank this individual as a Visionary (1 to 10 with 10 being the highest) and why?
	Yes or No	
	Yes or No	
	Yes or No	

What specific role will you ask these individuals to play?

The Positive Pessimist: Of course, just as important as the Visionary is what some call the Positive Pessimist.

So, what is a Positive Pessimist? Positive thinking people set their minds to believe a goal can be reached. The opposite of Optimistic thinking is Pessimistic thinking where people set their minds on all that can go wrong when trying to reach a goal. Positive Pessimists are people who set their mind to believe a goal <u>can</u> be reached, however, at the same time they set their minds to focus on and manage things that <u>can</u> go wrong in the process. A Positive Pessimist can help you keep your balance and ensure that you take nothing for granted and manage all aspects of your transformation process.

Name /Position/Title	Current Relationship?	How would you rank this individual as a Positive Pessimist (1 to 10 with 10 being the highest) and why?
	Yes or No	
	Yes or No	
	Yes or No	

What specific role will you ask these individuals to play?

The Implementation Specialist: All the visioning and planning in the world means nothing if you do not fully execute your action plan.

People who play the Implementer role are those who actually get things done. They are practical, efficient and well organized. An Implementer can help you turn your personal vision and Leadership development goals into executable plans. Implementers are also good finishers and have an eye for detail. They can help you detect missteps or omissions and help ensure that you adhere to your action plan milestones.

Name /Position/Title	Current Relationship?	How would you rank this individual as an Implementation Specialist (1 to 10 with 10 being the highest) and why?
	Yes or No	
	Yes or No	
	Yes or No	

What specific role will you ask these individuals to play?

The Emotional Supporter: We cannot stress the importance of this character type enough. Even the most skilled professional is only human, and life as a human can be challenging.

One of our most basic needs as human beings is real, authentic and meaningful connections with other people. Emotional support is about helping to lift someone to higher ground so he or she can see his or her way through the difficulty. With a respected and trusted Emotional Supporter on your team, you will have someone to rely on and to discuss your transformation progress. The connection can also help minimize some normal fears and suspicions.

Name /Position/Title	Current Relationship?	How would you rank this individual as an Emotional Supporter (1 to 10 with 10 being the highest) and why?
	Yes or No	
	Yes or No	
	Yes or No	

What specific role will you ask these individuals to play?

The Subject Matter Expert: Whatever your transformation path, there are people who know more than you do about the subject matter and the Leadership development needs you will be targeting.

There is no way to overestimate the importance of learning from the people around you. A Subject Matter Expert or Mentor can help you save time and ensure that you are on the right path. He or she can provide you practical insights and expertise in a specific subject, business area and Leadership style. During the transformation process, it is essential that you learn as much as possible from the experience of others.

Name /Position/Title	Current Relationship?	How would you rank this individual as a Subject Matter Expert (1 to 10 with 10 being the highest) and why?
	Yes or No	
	Yes or No	
	Yes or No	

What specific role will you ask these individuals to play?

> **The Old Hand:** Turning your Personal Leadership Vision into reality is enough of a challenge without making it harder for yourself by reinventing the wheel unnecessarily.

It is important for an aspiring Leader to grow and gain wisdom. Having someone who has "been there and done that" on your Support Team can be extremely beneficial. As a Leader, you must seek relevant wisdom and realize that time is a key factor in its acquisition. Albert Einstein once said, *"Wisdom is not a product of schooling but of the lifelong attempt to acquire it."* This is why those who are "older" or seasoned Leaders can be invaluable in supporting personal and professional transformations.

Name /Position/Title	Current Relationship?	How would you rank this individual as an Old Hand (1 to 10 with 10 being the highest) and why?
	Yes or No	
	Yes or No	
	Yes or No	

What specific role will you ask these individuals to play?

A Transformation Communications Strategy

A transformation communications strategy can keep all of your supporters and all of your stakeholders aligned with the primary target --- successfully completing your transformation from a good "Project Manager" to an "Outstanding Project Leader".

An effective communications strategy also provides structure to the interactions, gets the right details to the right people at the right time and prevents disparaging or false information from disseminating widely. Your communications strategy should be based on your actual needs. A Support Team that communicates efficiently performs effectively and will keep you informed of valuable reactions and feedback. They may also aid in executing your Transformation Action Plan.

Here is a template you should use to outline your communications strategy:

1. Current Situation/Background

Before you map out where you want your communications strategy to take you, you need to keep where you are "now" top of mind. The work you performed in section 2 can help here. The focus should be on what may have already been accomplished thus far from a communications point of view? How effectively have you previously communicated your career goals to others (management, co-workers and family members) and your professional development and career needs? What new communication vehicles are needed now to support your new transformation effort?

2. Overall Transformation Objectives

Your communications strategy should support the specific objectives included in your Transformation Action Plan. Use this section to list your "general" transformation targets. The work you performed in section 3 can help you here. You can then develop "specific" objectives to help achieve the actions needed to complete your Plan.

3. Communications Objectives

Use this section to list your communications objectives internal and external to your current workplace. Clear, specific, and measurable objectives are essential to the success of any communications strategy. Your objectives represent the impact you would like to see as a result of your communications with your Support Team and others.

4. Target Audience

Who do you want to primarily get your message across to in addition to your Support Team? Be as specific as possible:

Also, what about your Secondary Target Audience? – people of less importance who you wish to receive the communications messages, people who will also benefit from hearing your Personal Leadership Vision or people who might or can influence your target audience now or in the future.

5. Key Message per Target Audience

Develop specific messages for each Target Audience that can be easily understood and communicated, (i.e. "I want to…", "I plan to…" I will…").

Benefiting From Your Support Team and Communications Strategy

Remember, it is <u>your</u> Transformation Support Team and <u>your</u> communications strategy.

You must find the time to stay in contact with all the members of your Team and to keep your communications strategy current based on how your transformation process unfolds.

Also, be aware that the actual benefits gained from your Support Team will be dependent upon who you select as members of your team and your particular work situation and environment. The effectiveness of your Support Team communications strategy will depend on your audience, the quality of the message content you provide, the communication channels you use and the frequency of your contact.

PART II

EMBRACING THE 7-HABITS OF SUCCESSFUL PROJECT LEADERS

"Successful people are simply those with successful habits."

— Brian Tracy

✓ If the team members complain about too much information, explain that you are communicating across a wide spectrum and they are free to filter information, as long as they recognize it is their choice. If the client/sponsor complains, work to develop a communication schedule and format. Document the adjustment and ask them to approve the plan.

Why this is important?

What has been your experience?

Remember, *your thoughts become your actions*. When you make it a habit to monitor and get feedback on the quantity, effectiveness of your communications at all levels, and from all directions, you will always be in a better position to take the actions necessary to stay on course and avoid unnecessary pitfalls.

What are your thoughts and questions regarding this section?

I. Communicating Vertically, Horizontally and Often

As Project Leader, reporting to the Senior Vice President of Implementation, your new project is to assist the Global Rocket Company as they implement a new accounting and financial reporting system in six locations.

Outline a communications strategy for both your Project Team and your client which incorporates the essence of communicating vertically and horizontally.

6.

Planning as a Team, Executing as a Team and Winning as a Team

To have a great project team, there is no magic recipe for success. Our experience indicates that the combination of strong Leadership, effective communication and access to adequate resources contribute to productive collaboration. However, it all comes down to having team members who understand each other and work well together to achieve ultimate success. Having a Project Leader who has made it a habit of consistently fostering the right mix of planning, trust, ambition and encouragement among his or her team members is crucial to the execution and success of any type and size project.

THOUGHTS TO BE CONSIDERED

- ✓ As set forth in the Project Management Body of Knowledge (PMBOK®), your success as a Project Management Professional will depend on your skills, ability and knowledge in planning. However, since every project is unique in the problems that arise, the priorities and the environment in which it operates, successful Project Leaders realize that not involving the project team in the planning process, as much as possible, can lead to critical knowledge gaps as to the true project path.

Why this is important?

What has been your experience?

✓ As we all know, a good plan is dynamic, flexible and constructed to capture and track the actual project path as well as crystallizing the path in the minds of all team members. The key to fostering the right mix of planning, trust, ambition and encouragement is cultivating a sense of ownership.

Why this is important?

What has been your experience?

✓ By including the entire project team in the planning process, they will gain a personal sense of responsibility and ownership for the project and their individual tasks. If team members do not feel a personal sense of responsibility and ownership for the project, they will distance themselves from it at the first sign of trouble.

Why this is important?

What has been your experience?

✓ When all aspects of the project execution are recognized as valuable contributions to the team's overall success, all of your team members will feel that they are working together toward a common goal. As a result, when the project successfully crosses the finish line...it will be seen and will feel like a team win.

Why this is important?
What has been your experience?

As a habit, Outstanding Project Leaders not only value the idea of planning, executing and winning as a team, they also understand that it is their role to create a project environment which facilitates and encourages the maximum level of team involvement.

What are your thoughts and questions regarding this section?

II. Planning as a Team, Executing as a Team, Winning as a Team

In a not-for-profit environment with volunteers (no paid staff), you are asked to plan a fundraiser - a one day football jamboree where 5 high school teams play each other for one quarter each. There is one field to use. Teams do not need to stay overnight. You should consider parking, food, team rest and warm up areas, officials, crowd control, etc.

Outline how you would get everyone involved in this planning process.

Proactively and Fearlessly Managing Project and Resource Change

As a habit, successful Project Leaders control project and resource change by formalizing a carefully crafted change management process. This process should allow the project team to be empowered with a tool that applies a consistent rubric against every change no matter how small or simple it may seem. This kind of process ensures that every change is evaluated against the project objectives as a whole.

THOUGHTS TO BE CONSIDERED

- ✓ The most Outstanding Project Leaders understand the value of being proactive and fearless throughout the change management process.

Why this is important?

What has been your experience?

✓ Why does being proactive set you apart from other Project Leaders? Although many will not admit it, most Project Leaders think reactively and not proactively.

Why this is important?

What has been your experience?

✓ Being a proactive Leader means that instead of merely reacting to potential project changes as they happen, you consciously anticipate change-related events based on your intuition, calculation and experience.

Why this is important?

What has been your experience?

✓ Being proactive gives you the time and the opportunity to think through the options available and actions necessary to minimize project impact and risks.

Why this is important?

What has been your experience?

Now, when you combine being proactive with the courage and lack of fear to "do the right thing when it needs to be done," you will elevate your leadership presence, your influence and your team's level of success.

What are your thoughts and questions regarding this section?

III. Proactively and Fearlessly Managing Project and Resource Change

You are the Project Leader for a 3-month training project for an employee benefits company. The day before project kick-off the client calls to tell you the budget has been reduced by 15%, which equates to two employees. She also asks that you expand the training to include another 25 people, yet remain on the same schedule.

Explain how you would manage this situation.

Linking Risk to Common Objectives with a Compelling Vision

Vision is defined as the "power of discerning future conditions; shrewdness in planning and foresight". To some extent, we all have such power. Some of us are more inherently capable than others to look into the future...whether it is next year or next week...and plan for specific outcomes. However, the primary and ultimate challenge for most Project Management Professionals lies in how to articulate a compelling project vision. A vision that will "pave the road" that his or her project team will take to achieve the ultimate goal and a vision that successfully links project risk to the common objectives.

Thoughts to be Considered

- ✓ When a Project Leader confidently articulates a clear and compelling vision, he or she provides the project team members with the tools and knowledge to confidently move forward, execute with clarity and have a strong sense of the "risk versus reward" associated with the mission at hand.

Why this is important?

What has been your experience?

✓ The visioning process will also help you construct a Leadership action plan to help you guide your team successfully through each project phase and milestone.

Why this important?

What has been your experience?

✓ At the start of all major projects, most Outstanding Project Leaders make it a habit to construct a simple table linking or mapping all project objectives to potential risks and the controls that should be established to mitigate those risks …and make the road smoother.

Why this is important?

What has been your experience?

By clearly linking objectives, risks and controls, you and your project team will become more effective in gathering and maintaining documented evidence of the project's status and potential risks. This important "linkage" will also aid in the development of the team commitment and trust needed to achieve the ultimate project goal.

What are your thoughts and questions regarding this section?

IV. Linking Risk to Common Objectives with a Compelling Vision

You lead a team of ten that will develop a plan to convert an abandoned shopping mall into an office complex. This project will create thousands of jobs in an area that is suffering from lost industrial jobs.

Outline a vision that would be compelling and recognizes the risks and rewards for a successful conversion.

Approaching Problem Solving as a Creative and Analytical Process

As the Project Leader, you are not expected to have all the expertise to solve the multitude of challenges and problems that will inevitably surface during the execution of a complex project. However, you are expected to gather the right team of experts and to make sure that the best solutions are found for even the most difficult problems. To accomplish this important task, most successful Project Leaders start by ensuring that his or her team approaches the solution of major problems as both a creative and analytical process.

THOUGHTS TO BE CONSIDERED

- ✓ Approaching a problem in an analytical and logical manner is most common, but it normally leads to only a few or the most obvious solutions. Strategically attacking the problem solving process with more creative thinking usually leads to an array of possible solutions.

Why this is important?

What has been your experience?

- ✓ Creative thinking starts from the description of the problem and diverges to the identification and evaluation of all available and feasible solutions. A creative problem solving process requires all of your team's imagination and what many call "thinking out of the box".

Why this is important?

What has been your experience?

- ✓ Although the two ways of approaching a problem are different, they are linked because one complements the other. This is evident in the fact that all of the solutions that may result from the creative process must later be analyzed to determine an optimum solution based on the current set of circumstances and resources. Outstanding Project Leaders are open to assistance from the team in problem solving. They encourage brainstorming.

Why this is important?

What has been your experience?

✓ Outstanding Project Leaders are not afraid to involve key stakeholders. They will ask, "Which of these solutions do you think is best for you and your organization?" Successful Project Leaders do not fall back into comfort zones and "cookie cutter solutions"; instead, they seek and develop solutions that fit the project and yield the best results.

Why this is important?

What has been your experience?

As a habit, Outstanding Project Leaders take a creative and analytical approach to problem solving in order to generate better solutions.

What are your thoughts and questions regarding this section?

V. Approaching Problem Solving as a Creative and Analytical Process

A rural Hospital Network, four hospitals and six Outpatient Clinics, needs a new supply chain system to reduce inventory, better manage transportation costs, and provide critical supplies and pharmaceuticals around the clock. They have a warehouse but it is not centrally located.

Outline how you would approach solving this problem.

Maintaining Project Quality with Incremental Measures and Team Focus

One of the most difficult tasks of any project is measuring and managing project quality. Most Outstanding Project Leaders faithfully hold to the adage, *"If you can't measure it, you can't manage it"*. They also realize that when they employ incremental measures throughout the project lifecycle and maintain a team-wide focus on building quality into their project performance as well as into their products and services, they get the best results.

Thoughts to be Considered

- ✓ Quality is the totality of characteristics and features of a product, service or performance that affects its ability to satisfy the stated or inferred needs.

Why this is important?

What has been your experience?

✓ Of course, measuring quality performance and quality outcomes can be achieved with various techniques: including milestones, weighted steps, value of work done, physical percent complete, earned value and other measures. You should carefully choose which technique is the best fit for the project.

Why this is important?

What has been your experience?

✓ Project quality and performance can be tracked by any appropriate measure – cost, hours, quantities, schedule, percent complete and other measures.

Why this is important?

What has been your experience?

✓ Strategically establishing incremental milestones will allow you and your team to more closely monitor both project performance and quality. Early detection of performance issues could minimize the need for major corrective actions later.

Why this is important?
What has been your experience?

Most Outstanding Project Leaders know that it is difficult to effectively measure the quality of results during the project execution. Therefore, they make it a habit to keep their team focused on improving the overall quality of project performance – which results in the improved quality of project deliverables.

What are your thoughts and questions regarding this section?

VI. Maintaining Project Quality with Incremental Measures and Team Focus

Using the plan, you outlined for the conversion of the mall to an office complex (page 118):

List a set of project quality and incremental measures that you would establish to maintain team focus and ensure success.

Accepting Your Role as the Chief Confidence Builder

A project team may be comprised of staff members from the same department, multiple departments or even several different organizations. The most common set of characteristics of a good team member will generally include being skilled, experienced, dependable and motivated. However, during the project execution, what distinguishes "good" team members from outstanding team members is the level of confidence they have in themselves, in their ability to get the resources they need and in you…the Project Leader.

THOUGHTS TO BE CONSIDERED

- ✓ As a Leader first and a manager second, one of your most important roles as a project management professional is to cultivate and build your team's confidence. Your team members should have confidence in themselves as well as confidence in your leadership skills. In general, team members will judge the probability of future success based on past performance.

Why this is important?

What has been your experience?

- ✓ As you work with your team through project execution, you will build a record of accomplishment based on project successes and failures. If you maintain a good record of accomplishment, you will create a sense of optimism that future project challenges will also be successfully conquered. If you have a record of disappointment, your team may lose confidence in you and the probability of the project successfully achieving its goals

Why this is important?
What has been your experience?

- ✓ There are many tested principles that you can deploy to help build confidence in your team members. They include holding team members accountable for their actions, providing decision-making opportunities, delegating important tasks and providing stretch assignments.

Why this is important?
What has been your experience?

- ✓ People respond to "this is what we will do" better than, "maybe this will work". President John F. Kennedy, did not say, *"Let's try our best to put a man on the moon"*. Instead he said; *"We choose to go to the Moon in this decade and do the other things, not because they are easy, but because they are hard; because that goal will serve to organize and measure the best of our energies and skills, because that challenge is one that we are willing to accept, one we are unwilling to postpone, and one we intend to win."*

Why this is important?

What has been your experience?

Taking the time to know your team as individuals, and as a group, will help you decide what confidence-building actions are appropriate and when to deploy them. As a habit, the most Outstanding Project Leaders accept and understand their important role as the Chief Confidence Builder.

What are your thoughts and questions regarding this section?

VII. Accepting Your Role as the Chief Confidence Builder

One of your team members, a young analyst, made a calculation error that was discovered before it was presented to the client. You, the analyst and two senior team members worked very late to correct it.

What should you, as the Project Leader, do to help the analyst and the senior team members regain mutual confidence and recover from this mishap?

PART III

HARNESSING THE WISDOM OF OTHERS

"Wisdom is not a product of schooling but of the lifelong attempt to acquire it."

— Albert Einstein

PART III

HARNESSING THE WISDOM OF OTHERS

Harnessing the wisdom of others is a critical part of completing your transformation.

How often are you the "smartest person in the room"? Do not be too humble to admit the truth, there are times that you are and it is a great thing. People want to know what you know. You are the subject matter expert, the acknowledged leader or instructor; you would not be in the position that you are in today if you were not occasionally "the smartest person in the room".

However, most of the time, you will not fill that role and that is okay – all of us can benefit from the wisdom of others. True professional growth occurs when we "seek" advice from others, "listen" to them and "apply" what they share.

SEEKING

The first step is to seek advice or direction, to ask for knowledge to grow professionally. The answer to your quest comes in many forms.

Many people benefit from mentors – people who provide direction, advice, feedback and positive reinforcement. We both benefited from mentors at different stages of our careers and were fortunate enough to have mentored others. Mentoring is probably the most fulfilling role of Leadership.

If a mentor is not available or you are uncomfortable with that type of relationship, there are plenty of avenues to seek knowledge. Read. Any bookstore, library or on-line bookseller has a plethora of good Leadership books available. As you travel through this workbook, you will find quotes from people who write about leadership. Explore his or her writings and you will find someone who resonates with you.

Classes are valuable as well, especially when you can interact with the instructor or classmates. Find the forum that is best for you and pursue the knowledge.

LISTENING

"Knowledge speaks, but wisdom listens." Jimi Hendrix

When you ask for information, you have two roles; first, to explain what you want to know and why. The second role is to listen attentively and absorb the wisdom. When you do that, you will grow professionally. Limit your interruptions to important questions that clarify what you heard. If the mentor prefers dialogue, great, however the impetus is for you to listen and to learn.

Remember, no one has ever learned anything while he (or she) was talking.

APPLYING

The final step involves combining what you have learned with what you know and applying it. The more immediately you apply a new idea, strategy or tactic, the stronger it is in your memory and the more likely you are to use it again.

In this part of the workbook, you will have the opportunity to learn how to harness as well as apply the wisdom of others to deepen your Leadership vision and skills. The topics covered in Part III shed some contemporary insights on facilitating professional and career change as well as using your own personal strengths to identify and benefit from rewarding relationships. The topics include:

- *Thinking Like a Leader – Perspective and Learning;*
- *Understanding and Using the Power of Micro-Messaging;*
- *Identifying Career Critical Mentor Relationships; and*
- *Developing your "Mentor Ask" and Follow-up Strategy.*

The use of the word "harnessing" here is to emphasize the importance of not just strategically and professionally "attaching" yourself (and your career) to the constructive thoughts and experiences of others, but also for you to always "be in control' of the valuable advice and even more valuable relationships.

Let's get started.

12.

Thinking Like a Leader – Perspective and Learning

Every day a manager presents an idea to the CEO. He explains the benefits, the ROI, the long-term advantages and the potential revenue increase. He feels confident that he has provided a solution or great idea. The CEO nods, thinks for a minute before she asks, "How will this differentiate us from our competition?"

The manager is stunned; he did not anticipate that question. Why not? He thought like a manager or salesperson, not like a Leader. A Leader inspects ideas in a spherical or global manner, looking from many angles and asking questions to make sure the idea or plan is strong and thoroughly developed.

Project team members depend on their Project Manager to be a "Leader" and to act like their CEO. Outstanding Project Leaders have the skills and knowledge necessary to lead a team toward the successful completion of a project by inspecting project plans in a spherical manner, looking at stakeholders requirements from many angles, asking the right questions and providing the resources needed.

However, most Project Managers, who want to become Project Leaders, go about it the wrong way. Industry and academic experts all agree that in addition to exceptional project management skills, Outstanding Project Leaders in today's global, fast-paced business environment must gain a combination of insight and intellectual prowess, which allows them to:

- Visualize and manage the "big" project picture;
- Neutralize difficult project relationships and strengthen supportive relationships;
- Gain the trust of key shareholders and the favor of the doubters; and
- Navigate shifting business strategies in order to achieve the best possible outcomes.

Acquiring this type of Leadership foundation requires more than day-to-day project management experience and team building skills. It requires that you first build an intellectual foundation, which allows you to *"think like a Leader"*.

Successfully transforming from "Manager" to "Leader" starts with transforming how you think [about yourself, others, results, and organizations] and ends in the actions which arise from those thoughts.

Many Project Managers can recall a moment during the execution of a challenging project when something within them kicked in and said, *"It's time for me to step up and be a better project leader here."* Chances are "that moment" resulted in them making a very difficult decision; such as, saying "no" to an absurd request from a powerful stakeholder or persuading an executive sponsor to change project direction due to previously unknown and unsurmountable obstacles threatening to seriously affect cost and schedule. At that time, they seem to have gone through a transformation of some sort. They suddenly had a momentary glimpse of the "bigger" picture and, at least for a moment, they felt the vibes from their project team members acknowledging an Outstanding Leadership performance.

Most good Project Managers will have good Leadership moments. On occasion, they will appear to be Outstanding Leaders. However, the concept of transforming from a good Project Manager to an Outstanding Project Leader is a very challenging one.

Building the foundation required to successfully complete this transformation and maintain a consistent level of Leadership "thinking" requires:

- Broadening your perspective regarding Leadership;

- Gaining a visceral understanding of the role and mission of a "Manager" versus a "Leader"; and

- Learning how to "think like a Leader" by taking specific and personal actions that will strengthen your focus on how a leader thinks.

In this section, you will be presented with exercises to aid in helping you achieve both a broader perspective regarding Leadership and develop a plan of action to strengthen your focus on *"thinking like a Leader."*

This is not an overnight transformation; it evolves from repetition and a variety of opportunities to engage the concept. However, transforming your thinking is a vital part of your transformation from a good Project Manager to an Outstanding Project Leader.

Okay. Let's take the first steps towards building a foundation, which will support your transformation.

Understanding the Role and Mission of Management versus Leadership

Here are three contemporary perspectives to help broadened your thinking of the role and mission of a "Manager". Take a moment to think about and document your current perspective on how a "Leader" differs.

Manager vs. Leader Personality

A managerial culture emphasizes rationality and control. Whether his or her energies are directed toward goals, resources, organization structures or people, a manager is a problem solver. The manager asks: "What problems have to be solved, and what are the best ways to achieve results so that people will continue to contribute to this organization?" From this perspective, management is simply a practical effort to direct affairs. To fulfill his or her task, a manager requires that many people operate efficiently at different levels of status and responsibility. It takes neither genius nor heroism to be a manager, but rather persistence, tough-mindedness, hard work, intelligence, analytical ability, and perhaps most important, tolerance and goodwill.

What is your current perspective on a Leader's Personality?

Conceptions of Work

Managers tend to view work as an enabling process involving some combination of people and ideas interacting to establish strategies and make decisions. They help the process along by calculating the interests in opposition, planning when controversial issues should surface, and reducing tensions. In this enabling process, managers' tactics appear flexible: on one hand, they negotiate and bargain; on the other, they use rewards, punishments, and other forms of coercion.

What is your current perspective on a Leader's Conception of Work?

Relations with Others

Managers prefer to work with people; they avoid solitary activity because it makes them anxious. Several years ago, a group of researchers directed studies on the psychological aspects of careers. The need to seek out others with whom to work and collaborate seemed to stand out as an important characteristic of managers. When asked, for example, to write imaginative stories in response to a picture showing a single figure (a boy contemplating a violin or a man silhouetted in a state of reflection), managers populated their stories with people.

What is your current perspective on a Leader's relations with others?

MANAGER VERSUS LEADER

This exercise is designed for you to gain some additional insight into how you currently think of being a "Manager" versus being a "Leader". Read each of the seven statements in the *left* column regarding what a manager does. Then, add your *thoughts* in the *middle* column on what you think a Leader does. In the *right* column, indicate what drives your thinking in this direction. Our *thoughts* are on the last page in this section.

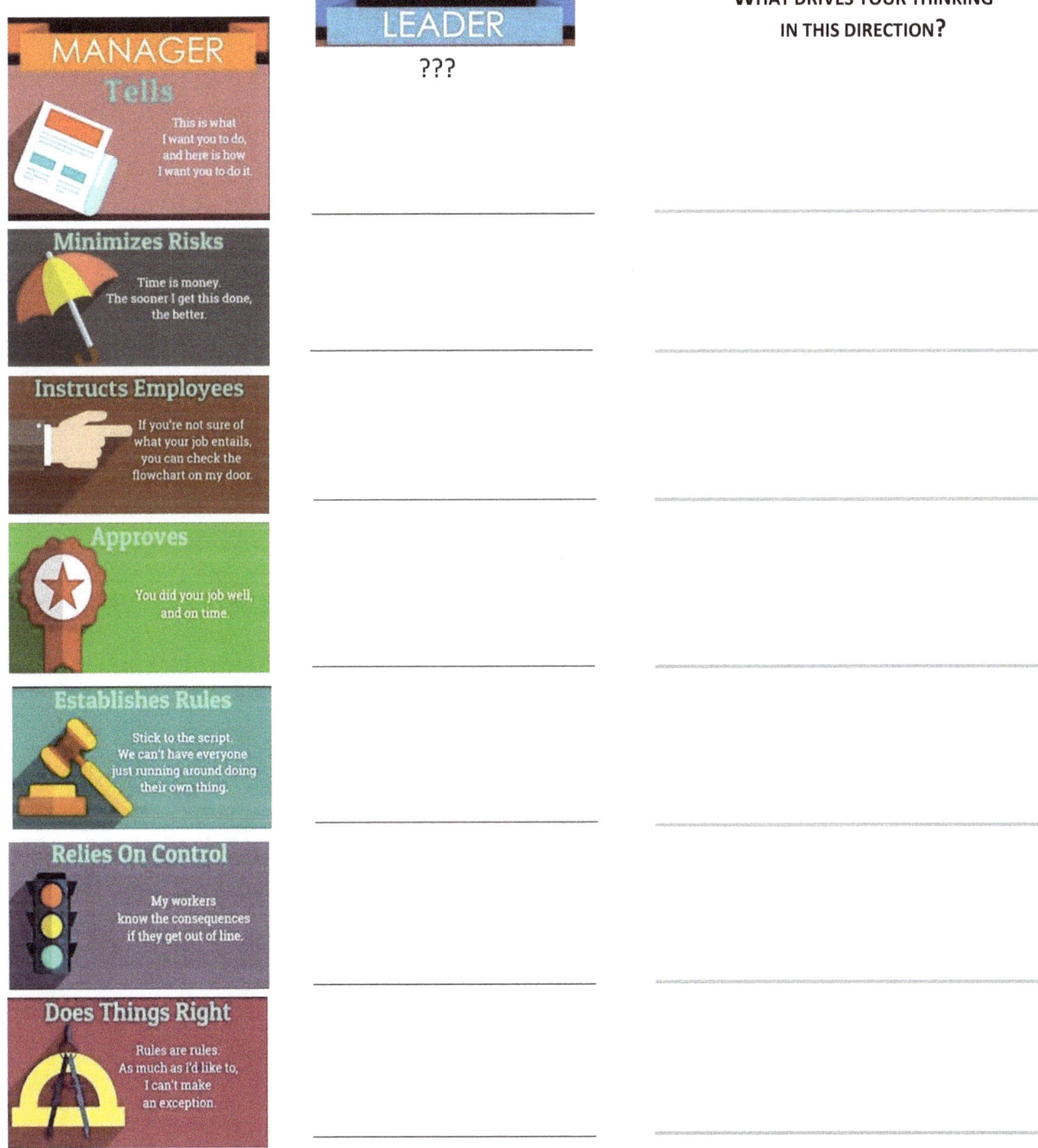

NOTES

Learning How to "Think Like a Leader"

One of the most effective ways to learn how to "think like a Leader" is to take specific and personal actions to strengthen your focus on how a Leader thinks. Consider each of the seven actions below and outline how you will approach taking these *actions* during your transformation.

1. I will cultivate strategic thinking.

Strategic thinkers can simplify the difficult, prepare for uncertainties and reduce the margin of errors--all because they have a plan. Strategic thinking makes you a better planner, which is how you move easily from where you are today to where you want to be tomorrow.

I will approach taking this action during my transformation by:

2. I will engage in inquisitive thinking.

Outstanding Leaders spend their time questioning everything they *know* and everything they *don't know*. When you question, you gain knowledge, and when you gain knowledge, you have impact. To be impactful, you have to question what everyone else is taking for granted. Taking this approach alone can give you a leg up on innovation and creativity.

I will approach taking this action during my transformation by:

3. I will explore big-picture thinking.

Big-picture thinkers are always ready to see things that other people cannot see; they are able to size up a situation and take all the variables into account. Once you can connect the dots like no one else, you will always be prepared to seize an opportunity when the time is right.

> *I will approach taking this action during my transformation by:*

4. I will harness focused thinking.

Focused thinking shuts out interruptions and interference, allowing you to concentrate with clarity. When you can focus your thinking, you are able to bring clarity to challenges, targets, and results.

> *I will approach taking this action during my transformation by:*

5. I will utilize risk-oriented thinking.

Outstanding Leaders think big and dream bigger than most. When you learn how to push the envelope and dare to go where no one else has even looked yet, you will be admired as a risk taker, someone who dares to gamble--and because you dare more, you will gain more.

I will approach taking this action during my transformation by:

6. I will rely on collaborative thinking.

Collaborative thinkers like to hear what other people are thinking so they can expand their own ideas. As much as we like to think we know it all, the best kind of thinking--the kind that brings the greatest return--is not done solo but is shared.

I will approach taking this action during my transformation by:

7. I will practice reflective thinking.

Take the time to reflect before you act, listen before you speak, understand before you respond and engage your compassion before you react. When you take the time to reflect, you gain perspective. It allows you the bandwidth to see what is truly going on without being emotionally charged. Reflective thinking enables you to distance yourself, so you can see things with a new pair of eyes.

I will approach taking this action during my transformation by:

MANAGER VERSUS LEADER

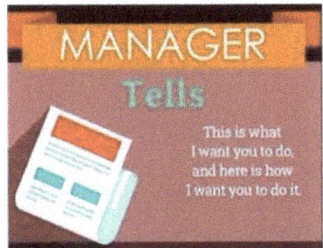
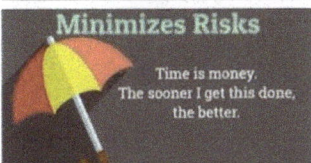
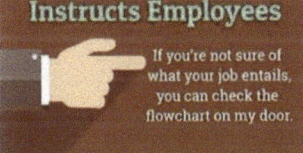

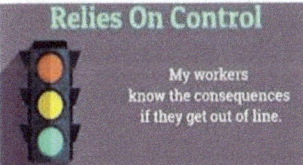
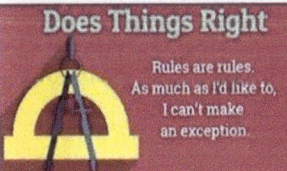
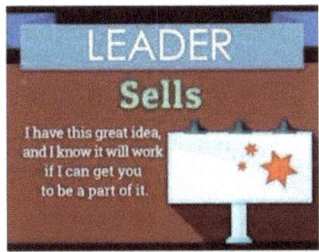
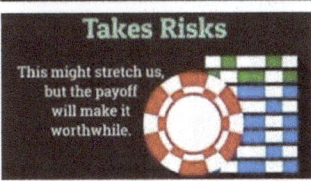
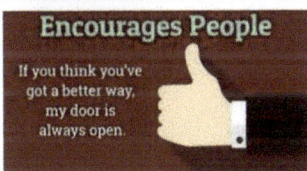

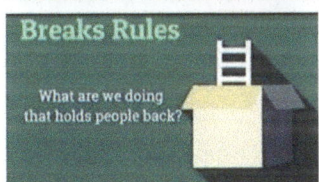

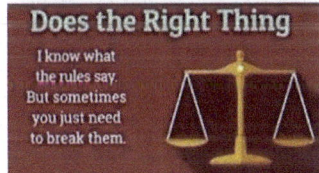

WHAT DRIVES A LEADER'S THINKING

- Outstanding Leaders are outstanding salespeople. They must sell their vision and their ideas. Their ability to influence and persuade others is critical to their success.

- Outstanding Leaders take carefully calculated risks, while accepting that failure is a byproduct of success and innovation.

- Outstanding Leaders know that "how" and "what" they communicate affects performance.

- Outstanding Leaders do not leave motivation to chance. They know that the slightest mishap, whether internal or external, can bring momentum to a grinding halt.

- Outstanding Leaders know that some rules are made to be broken. They work hard at figuring out which ones they are and what it means to break them.

- Outstanding Leaders know that Inspiring trust is critical. People are not willing to recognize someone as their Leader unless they trust them, not just intellectually, but ethically and morally as well.

- As strategic thinkers, Outstanding Leaders are concerned with "doing the right things". They view "doing the right thing" as making choices among possibilities in favor of collective wisdom.

NOTES

13.

Understanding and Using the Power of Micro-messaging

You have done it and it has also been done unto you. You shake a person's hand, but barely make eye contact. You sit in on a colleague's presentation and repeatedly glance down at your watch. You "listen" while pecking away at a text message. It turns out that these seemingly unrelated acts of disrespect have three things in common: Micro-messaging, Micro-inequities and Micro-affirmations. Their impact goes far deeper than what meets the eye.

In 1973, MIT researcher Mary Rowe, PhD coined these terms. The term "micro messages" is commonly defined as:

- Subtle slights and snubs that devalues others;
- Instances of subtle interactions which are perceived as imbalances of human actions; and
- Communicating who is in the inner circle and who is not.

Embedded within the most common "micro-messages" are indirect offenses that can demoralize others, called "micro-inequities". Micro-inequities are the seemingly harmless messages of devaluation. They are a subset of the estimated 2,000 to 4,000 micro-messages that individuals send each day. Researchers cite that in the space of a one-minute conversation, each individual will send between 40 and 50 micro-messages to one another. These "small bits of meaning" occupy a continuum. "Positive" micro-affirmations are on one-end and "negative" micro-inequities on the other end. In the case of the latter, researchers contend that these negative micro-messages are rooted in powerful biases that are often subconscious.

In addition, while that is true of micro-inequities, we can have "magic" in a sincere micro-moment. That is because when, as a Leader, your micro-message repertoire is full with messages that are positive, affirming and appreciative, you do two things:

1) Build trust and strengthen relationships among members of your team and others in your organization; and

2) Provide a compelling model of appreciation and respect for others to follow. Leaders, who master "the how in the what," and the "magic" within the micro-moments, are rewarded with engaged teams and better performance.

You too can tap into the power of micro-messaging. In this section, you will be enlightened on the meaning, characteristics and impacts of the micro-messaging concept.

The insights and questions are intended to provide you some important background, challenge your thoughts and explore possible actions you can take to deploy this unique and intriguing component of Leadership to your advantage.

This section will briefly:

- Provide an overview of the Micro-inequity and Micro-affirmation framework; and
- Present some important questions you should consider regarding Micro-inequity and Micro-affirmation.

We hope that you will take it from here and determine how to both be observant of the concept and how to utilize it to your advantage during and after your transformation.

The Micro-inequity and Micro-affirmation Framework

What are Micro-inequities?

"Micro-inequities are brief and commonplace daily verbal, behavioral, and environmental indignities, whether intentional or unintentional, that communicate hostile, derogatory, or negative racial, gender, sexual- orientation, and religious slights and insults to the target person or group." (Rowe, 2008)

What are the Characteristics of Micro-inequities?

- Constantly and continuously experienced.
- Often committed unknowingly by well-intentioned people.
    - Dismissed as innocent or innocuous.
    - Gaps between our lived experiences.
- Micro-level appearances of enduring institutional and systemic imbalances of privilege and power.
- Can have verbal, nonverbal or environment appearances.

What are the Typical Forms of Micro-inequities?

- Micro-insults:
    - Often unconscious communications that convey rudeness and insensitivity and demean a person's identity.
- Micro-assault:
    - Often conscious, explicit derogations characterized primarily by a violent verbal, nonverbal or environmental attack meant to hurt the intended victim through name-calling, avoidant behavior or purposeful discriminatory actions.
- Micro-invalidation
    - Often unconscious communications that exclude, negate, or nullify the psychological thought, feelings, or experiential realities of people who experience micro-aggressions.

What are some examples of Micro-inequities?

- Checking emails or texting during a face-to-face conversation.
- Consistently mispronouncing a team member's name.
- Interrupting a team member in mid-sentence.
- Making eye contact only with males (or females) while talking to a group containing both males and females.
- Taking more questions from male team members than female (or Vis versa).
- Confusing a team member of a certain ethnicity with another person of the same ethnicity.
- Rolling your eyes.
- Sighing loudly.
- Raising your voice, even though a team member has no difficulties hearing you.
- Mentioning the achievements of some team members at a meeting but not others whose achievements are equally relevant.
- Consistently ignoring a team member's emails for no good reason.
- Only reading half of a team member's email and then asking him or her about the content later.
- Making jokes aimed at certain groups.

What is the Impact of Micro-inequities?

- Both committers and targets are impacted.
- Psychological and physical consequences.
- Unnecessary individual and team stress.

What are your thoughts at this point regarding your understanding of Micro-inequities?

What are Micro-affirmations?

"Apparently small acts, which are often transient and hard-to-see, events that are public and private, often unconscious but very effective, which occur wherever people wish to help others to succeed." (Rowe, 2008)

What are some examples of Micro-affirmations?

- When a team member tells you they feel they have been targeted because of their identity, you believe them.
- Acknowledging that a micro-inequity may have occurred.
- Visibly confronting inequitable, hostile, or biased behavior.
- Stopping to ask for a team member's opinion or contribution who has not had a chance to speak (in a group setting or during a meeting).
- An advertisement for an event includes an invitation to contact you if a team member with a disability needs an accommodation to attend.

What is the Impact of Micro-affirmations?

- Consistent, appropriate affirmation of others can spread from one person to another.
- Many micro-inequities are not conscious, but affirming others can become a conscious as well as unconscious practice that prevents unconscious slights.
- Acknowledging the existence and experience of micro-inequities allows the individual to confirm that they did not imagine these small, demeaning acts.

Questions to Consider

How are micro-inequities being experienced by various groups within your team(s) and organization?

How can an awareness of micro-inequities improve your team's ability to work together and improve their performance?

How can effective mentoring practices and micro-affirmations work to counteract the effects of micro-inequities within your team(s) and organization?

How do you plan to address micro-inequities immediately or shortly after they occur in the future?

14.

Identifying Career Critical Mentor Relationships

Selecting the correct mentor, at any point in your career, is a critical decision. You have to choose the correct person, who will provide the correct message, at the correct time. Compare it to hitting a major league fastball – you have to calculate speed, location and movement in less than four tenths of a second, and then swing on the proper path. The good news for you is that you get more than four tenths of a second to decide.

Determining the ideal mentor is both simple and complex. Simple because it must be someone who is successful and respected in your field or a closely related field and he or she can help you grow professionally. Then, it becomes complex.

First, you need a *relationship* to form the foundation. A common employer, a professional connection, a common professional organization or a friendship, can form this relationship or it can be a matter of being in the right place at the right time.

Second, the mentor must have the *time* and *interest* in assisting you. Time is a valuable commodity and if your potential mentor does not have the time to assist you, you need to know right away. Everyone has multitudes of time commitments and restraints to balance; mentoring requires time and not everyone has time to invest in you. He or she must have a genuine interest in assisting you.

Finally, what do you want to learn from your mentor? This element of the equation is equal to each of the other elements. Until you know the answer, you cannot move forward. Becoming a Leader requires knowledge, decision-making, strategic thinking, delegation, communication, people skills, time management, flexibility, management skills, patience, problem recognition, problem solving, planning, analytical skills and a host of other talents and skills. *What is your most important development need right now?*

This section will help you identify your critical career needs, potential mentor candidates and the type of mentor relationships that would be best for you.

Let's dig deeper into identifying mentor relationships and be prepared to hit a home run.

"The delicate balance of mentoring someone is not creating them in your own image, but giving them the opportunity to create themselves."

— Steven Spielberg

A. Identifying Your Career Critical Needs

What changes do you need to make to transform from "Manager" to "Leader"? The only incorrect answer is "nothing".

We all need to change (improve) in some area of our professional and personal lives. The most successful people recognize that fact, make an honest assessment and invest the time improving. Outstanding Project Leaders continue to improve in some facet of their lives continuously.

Close your eyes and imagine the Leader that you want to become – see yourself in that role. How do you look, act, speak and react? How do people react to you? How does it feel to visualize you in that role?

What changes do you need to make to become that person?

To help you feel less self-conscious about this, we have listed some of the changes that you might need to make below.

- Become an expert in your field;
- Improve your people skills;
- Obtain professional certifications;
- Understand interaction and interdependence with a project and an organization;
- Be more open to new ideas; and
- Listen more.

Now, it's your turn, the next page is blank, it is your canvas – where you should list the changes you need to make to become the Leader you want to become.

"There is one quality that one must possess to win, and that is definiteness of purpose, the knowledge of what one wants, and a burning desire to possess it." — **Napoleon Hill**

Your Career Critical Needs Today

What I Need	Why?
1.	
2.	
3.	
4.	
5.	
6.	
7.	

B. Identifying Potential Mentors

You have identified the changes you want to make; the next step is to identify someone who can help you during your transformation journey. Yes, you need someone to help you. Mentors provide advice, direction, and education. He or she will also shorten the learning curve.

We have both benefitted from mentors and have mentored; each relationship was unique and provided a positive result. Before Google, the most important source of information for many professionals was access to an expert. Someone who could say, *"You might have better results if you tried this way"* or *"I've made that mistake, here is what I learned"* or *"Let's discuss the best way for you to approach that situation"*.

The virtual communications age has created many more opportunities for mentoring. No longer does he or she have to be in the same company, city or region. You are not limited to breakfast, lunch or a physical meeting. You can schedule a call, exchange emails and documents even send a text message.

Is there someone whose professionalism and accomplishments make you say, "I can do that too?" Without geographic restraints, imagine how can you benefit from a mentoring relationship with this person?

Take a few minutes to think of seven people that you feel could provide advice and counsel as you improve your Leadership skills and thinking.

Name(s)	I Know this Person (X)	I Do Not Know This Person (X)	I Know Someone Who Might Know This Person (X)
1.			
2.			
3.			
4.			
5.			
6.			
7.			

Every young student has different teachers each year – because each teacher specializes in teaching a certain set of skills, then passing the students to the next teacher who builds on that foundation. Mentoring is similar – you need different advice at various stages of your career.

We will re-visit your list and discuss this topic further in Part IV.

"Mentoring is a brain to pick, an ear to listen, and a push in the right direction."

— John C. Crosby

C. Matching Your Mentor Relationships

What do you want to learn from your mentor?

Fortunately, when you determine what you need to learn, you are ready for a mentor and Buddha's statement will come to fruition: *"When the student is ready, the teacher will appear."*

What are your most important developmental needs right now and which person on your list is the best person to help you during your transformation?

Name(s) (From Page 157)	Rank the List Based of Best Match	Reason for the Ranking
1.		
2.		
3.		
4.		
5.		
6.		
7.		

"If I have seen further it is by standing on the shoulders of giants." — Isaac Newton

NOTES

15.

Developing Your "Mentor Ask" and Follow-up Strategy

You have identified your potential mentors, now you need to approach her or him and ask for help. It should not seem painful, if you prepare. Preparation is an important Leadership quality, the more important the issue, the more prepared you need to be. This is an important step for you; you need to prepare.

Athletes are trained to visualize successful outcomes, then practice to create them. You can mirror that by thinking about what you want to say, how you want to say it and anticipate questions - then receive the answer you want to hear: "Yes, I would be thrilled to help you!"

Let's get you to "Yes!"

"The best way a mentor can prepare another leader is to expose him or her to other great people."

— John C. Maxwell

A. Developing a Mentor "Ask" Format

"Hi, I was wondering if you would be my mentor for the next few months, it would mean a lot to me and help me advance my career."

While that is your thought, it sounds one sided (what *you* can do for *me*) and it does not sound confident or professional. This is not a date or an offer to attend a meeting/event. It is an important step in your transformation to become a more effective Leader.

Before you can ask a mentor for help, you must know what you want from the relationship.

Take a few minutes to reflect on these questions:

- What do you want from this relationship?

- Why this person can help you? (i.e. *honest, candid, well-positioned*)

- How do you envision the relationship? (i.e. *a good start, good sounding board*)

NOTES

B. Constructing a Mentor "Ask" Message

It is time to convert your thoughts into a succinct message - a message that is comfortable and states your case. It will not be your opening statement; it is the heart of your message. It needs to be comfortable to say, flexible (not a robotic, rehearsed declaration), sincere and professional.

There is no magic formula; however, you can construct your thoughts into a coherent message. This is not a monologue but a discussion. Your goals are to initiate the discussion and guide it towards the important questions you want to ask.

Here are a few ideas of what you might use as conversation starters:

- I wanted to talk with you about something very important and I hope you can provide me advice. I have observed how you handle some difficult situations (name one or two) and wish I could do that.

- What was your career path?

- I want to become recognized and rewarded for being an Outstanding Project Leader. What advice can you offer to get me on the right path?

- Did you have anyone provide you a boost in your career?

- Have you noticed anything that would prevent me from advancing my skills to those of an Outstanding Project Leader?

- I feel like I am ready to take the next step in my career; I am not sure how to do that. Do you have any advice?

Take your thoughts regarding critical career needs from Part A of this section and develop a few mentor conversation starters on the following page.

CONSTRUCTING YOUR MENTOR CONVERSATION STARTERS

Starter #1

Starter #2

Starter #3

C. How and When to Approach each Mentor Candidate

You now know what you want to say and you have a few conversation starters to get past the butterflies orbiting your stomach. How and when do you do it?

You do not need to find a romantic setting, purchase flowers or champagne or hire a violinist. What you want is a private setting with limited interruptions. If your potential mentor works at the same place, you definitely want a meeting, not a conversation in the cafeteria or lobby, where interruptions are guaranteed. If it is someone outside of your workplace, then a more casual setting might help you relax.

Once you are set, use one of the conversation starters to create a discussion, and then do the most important thing you can – LISTEN. *Is she enthusiastic when she answers? Does he give thoughtful, sincere answers or are his answers trite?*

Because you are asking someone you feel comfortable with, you will know when to "pop the question". Do not stick to a script, speak from the heart and ask her to mentor you, then stop speaking. Wait for a response.

The most likely responses are:

- "Yes"
- "How do you see this occurring?"
- "Why me?" or
- "I am flattered you think of me that way."

In the unlikely event the answer is no, ask if there is a reason for "No." It might be a busy schedule, personal conflict or discomfort.

After you remember to exhale, the two of you can discuss the process and details.

Remember this is not a one-sided relationship, it is a partnership and both parties need flexibility and clear goals.

Congratulations! You identified and secured a mentor.

You are on the Transformation Superhighway!

PART IV

DEVELOPING AND SYSTEMATICALLY EXECUTING YOUR ACTION PLAN

"Without execution, 'vision' is just another word for hallucination".

— Mark V. Hurd, CEO of Oracle Corporation

PART IV

DEVELOPING AND SYSTEMATICALLY EXECUTING YOUR ACTION PLAN

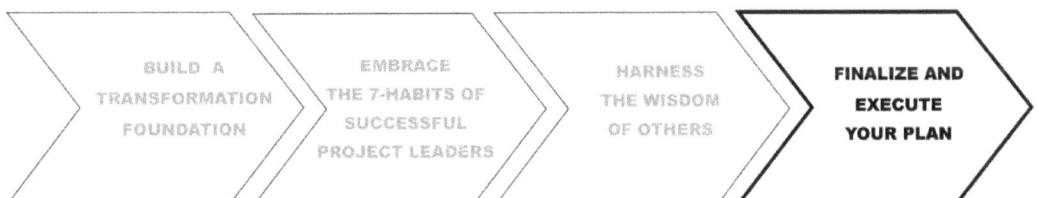

Now is the time to develop, finalize and systematically execute your action plan.

You are in the position to accelerate your transformation. Like astronauts, you have been through training, simulations and rehearsal, now it is time to strap into the command module and finish the preflight checklist. Time to launch!

In this section, you will accomplish four important tasks that will position you for transformation success:

- Construct Your Transformation Health Insight Summary (THIS);
- Understand the Action Planning Process and Determine Your Planning Guidelines;
- Develop Your Transformation Action Plan; and
- Finalize Your Action Plan.

How many people do you know who delay finding a new job or starting a new venture because "things are not quite right?" They are still waiting, aren't they? You cannot develop a perfect action plan. However, you can develop a plan that is ideal for you and a plan you can successfully execute. Yes, you might revise it as you go along, but you cannot wait for perfection, you must act now. Executing the ten actions within your *Transformation Action Plan* requires a significant amount of energy and a major dose of reality. You would not begin marathon training by running 26.2 miles the first day. You would increase your mileage each week for months to prepare for the grueling event. This principal applies to your action plan development and execution.

A plan is just a "to do" list unless you "act" on it. Your success, your future and your sense of accomplishment hinge on developing and executing your action plan.

It's time to turn the page and to act.

"You were born to win, but to be a winner, you must plan to win, prepare to win, and expect to win."

— Zig Ziglar

Constructing Your Transformation Health Insight Summary

The first step in determining the optimum guidelines to use in your Transformation Action Plan development is to reflect back on and summarize the awesome work you have done to this point.

Each of the twenty-nine exercises you have completed and the dozens of thought provoking questions you have responded to have differed. However, when combined, they have "jump started" your transformation through intense personal and professional exploration. The work that you have done has highlighted and revealed your professional skills, strengths, gaps, perceptions, perspectives and experiences while simultaneously considering the groups, organizations and work culture to which you currently belong.

Using the exercises you have completed in the previous 15 sections and the Leadership growth you have gained thus far, you will now be asked to construct your *Transformation Health Insight Summary (THIS)*.

An analysis of your "THIS" will help determine the optimum action planning approach. The optimum approach must include what we call "*Transformation Harvesting Actions and Thoughts (THAT)*". You have invested a tremendous amount of time, energy and resources in your education, training and career so far. It is now time to *harvest* the fruits of your investment by successfully executing the actions and Leadership thoughts included in your Transformation Action Plan.

Okay. Let's summarize the work you have done to this point and the insights gained thus far.

"It's important to have a sound idea, but the really important thing is the implementation."

— Wilbur Ross, United States Secretary of Commerce

Transformation Health Insight Summary

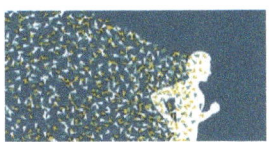

Based on the insights gained from the exercises you have finished, populate the following illustrations as instructed. Together the set of illustrations will comprise your *Transformation Health Insight Summary (THIS)*.

Leadership Performance Assessment (LPA)

Attribute	Your Score *(From Page 57)*
Leads by Example	
Visionary	
Strong Communicator	
Trustworthy	
Calm	
Open-Minded	
Professional	
Clear Thinking	
Nurturing	
Supportive	
Total *(Sum of Numbers in the Boxes Above)*	

Your LPA Index

LPA Index = Total/10

Leadership Attribute Analysis and Improvement (LAAI)

Number of Leading Attributes
(From page 58)

Number of Leading Attributes

Number of Lagging Attributes
(From page 58)

Number of Lagging Attributes

Project Leadership Assessment (PLA)

Exercise	Enter the Number (1,2 or 3) which Corresponds with Your Score	
Organizational Alignment **1** (35 or above) **2** (22-27) **3** (21 or less)		*(From Page 70)*
Leadership Type **1** (35 or above) **2** (22-27) **3** (21 or less)		*(From Page 72)*
Leadership Behaviors **1** (35 or above) **2** (22-27) **3** (21 or less)		*(From Page 74)*
Leadership Resilience **1** (35 or above) **2** (22-27) **3** (21 or less)		*(From Page 76)*
Project Leader Development Alignment **1** (35 or above) **2** (22-27) **3** (21 or less)		*(From Page 78)*
	Total *(Sum of Numbers in the Boxes Above)*	**Your PLA Index**

PLA Index = Total/5

Transformation Journey Approach (TJA)

Review Your Responses to Questions in Section 3	Indicate if You are Satisfied with Your Answers Enter "1" for YES or "0" for NO
How would you describe what you are about to do, why are you doing it and what benefits do you wish to gain? *(From page 80)*	
How will you create a sense of urgency within yourself and maintain it throughout the transformation process? *(From page 80)*	
How will you approach putting your Transformation Support Team in place? *(From page 81)*	
How will you engage your key stakeholders - particularly managers, family and friends - in helping you make this transformation happen? *(From page 81)*	
How will you approach "over-communicating" your Personal Leadership Vision? *(From page 82)*	
How will you approach the removal of obstacles to your transformation success? *(From page 82)*	

Review Your Responses to Questions in Section 3	Indicate if You are Satisfied with Your Answers Enter "1" for YES or "0" for NO
Are you comfortable with cultivating new ideas, learning from others and changing personal and professional behaviors? *(From page 83)*	
How will you go about realigning relationships (if required) to enable your vision and support your transformation? *(From page 83)*	
Why do you believe that you can upgrade your Leadership skills to the extent required to successfully complete your transformation? *(From page 84)*	
Total *(Sum of Numbers in the "nine" Boxes)*	
	Your TJA Index

TJA Index = Total/9

Transformation Support Team (TST)

How many potential Support Team Member names did you include for each of the following Character Types in Section 4 (Pages 89 – 94)	Enter "0" if no name was entered Enter "1" if one or more names were entered
The Visionary *(From page 89)*	
The Positive Pessimist *(From page 90)*	
The Implementation Specialist *(From page 91)*	
The Emotional Supporter *(From page 92)*	
The Subject Matter Expert *(From page 93)*	
The Old Hand *(From page 94)*	
Total *(Sum of Numbers in the "six" Boxes)*	
	Your TST Index

TJT Index = Total/6

17.

The Action Planning Process and Determining Your Guidelines

The *Transformation Action Planning Process* starts with a pre-determined set of ten transformational actions. This set of targeted actions establishes the baseline and directs the focus required to support the completion of your transformation. The action planning process involves the following steps:

STEP 1	STEP 2	STEP 3
✓ Determine the Appropriate Action Planning Guidelines.	✓ Follow the appropriate Guidelines to Develop a Set of Specific Objectives and Milestones for Each of the Ten Actions in the Action Planning Outline.	✓ Finalize Your Plan.

The set of targeted actions are embedded within the *Action Planning Outline* in section 18. The action planning process results in the completion of a comprehensive transformation plan. It involves requiring you to thoughtfully develop and insert a set of specific objectives and milestones in the outline to complete your plan. Successfully executing your plan and the ten actions should be your primary focus during the remainder of your transformation journey.

Based on our experience working with coaching clients over the years, we have been able to construct an *"Action Planning Guide"* to assist clients in the successful completion of this important phase of the program. The explicit, but very important, guidance is based on where we believe you are currently in your transformation process as determined by the insight gained from the *Transformation Health Insight Summary* constructed in the previous section.

Action Planning Outline

Determining the appropriate guidelines will help you to decide what objectives you should and can accomplish over the next three to six months.

Remember, your objectives are "structured steps" that you must make in order to achieve the actions in your transformation plan. A thoughtful and structured plan will help you maintain a balanced life while completing the transformation process. One of the primary advantages of the DUPP Program is that it is self-paced learning, ideal for balancing family, work and life. The flexibility allows you to commit the time you currently have available.

The most important aspect of a good transformation plan is that the objectives are realistic, achievable and SMART. By SMART, we mean:

Specific	All of the objectives associated with each action should be clearly defined.
Measurable/Milestones	You should quantify each objective with measurements and milestones, so you know when you have achieved it.
Attainable	You should thoughtfully establish the scope and milestones associated with all objectives to ensure that they are attainable within the planned timeframe.
Relevant	Your objectives should fit within a well-thought-out plan and are relevant to achieving your ultimate goal --- becoming an Outstanding Project Leader.
Timely	You should set a realistic date by which your objectives and the overarching actions can be achieved.

The remainder of this section will assist you in determining the guidelines you should use to thoughtfully develop a set of specific objectives and milestones for each targeted action in the planning outline.

You are getting close to being in the position to finalize your plan and beginning the final phase of your transition --- executing your Transformation Action Plan and becoming recognized as an Outstanding Project Leader.

Determining Your Action Planning Guidelines

To determine the "Guidelines" which can best assist you in the development of your Transformation Action Plan, populate the illustration below and then, simply follow the decision tree to determine your action planning guidelines.

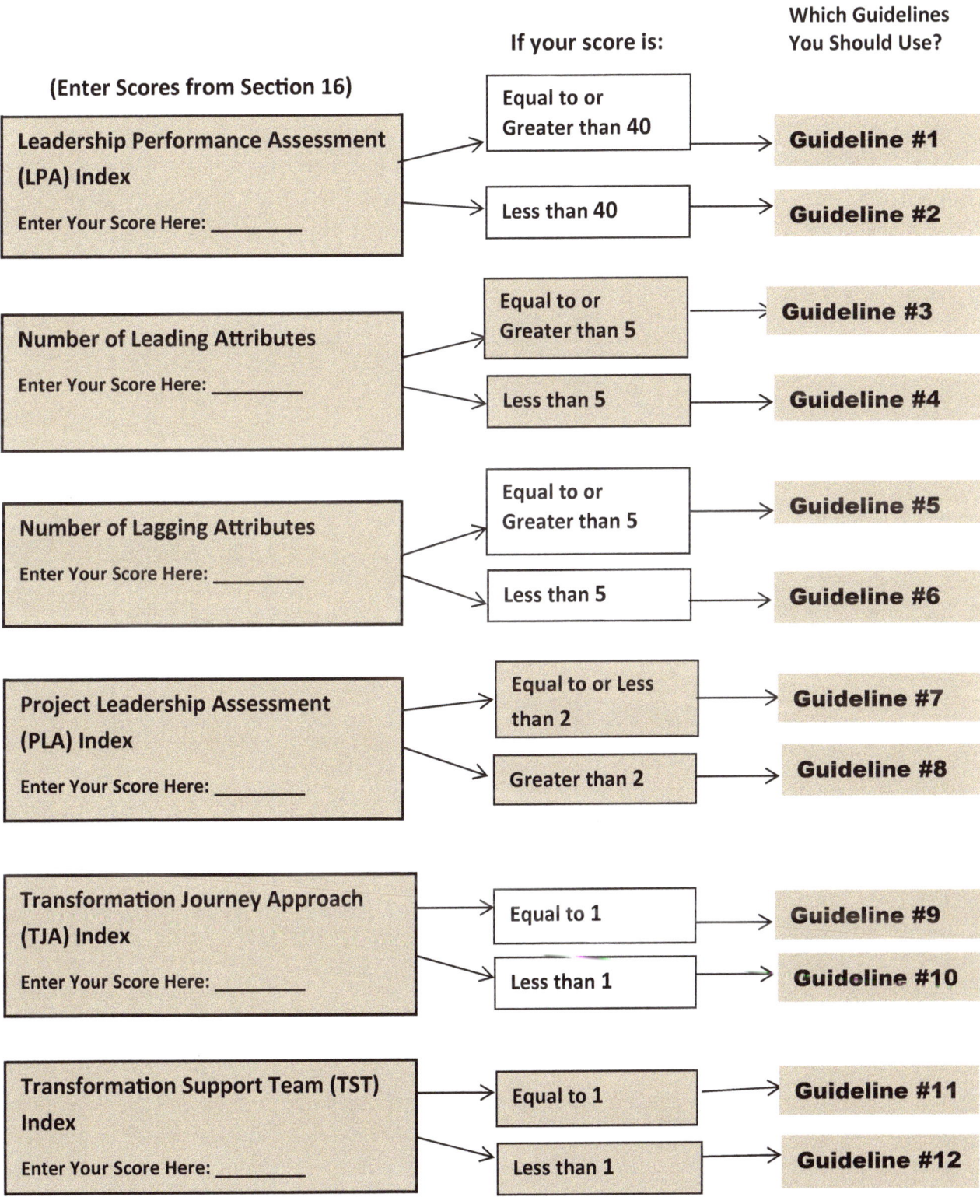

Action Planning Guidance

The Action Planning Guidance is uniquely constructed to assist you in the successful completion of this important phase of the program. The specific guidance is determined by factors from the *Transformation Health Insight Summary* constructed in the previous section. The ten *"Targeted Actions"* and the *"Suggested Guidelines"* will help direct your focus on what you must do to complete your transformation. The guidance will assist you in developing a set of specific objectives and milestones for each of the actions in the planning outline. The chart below shows the relationship between the *"Guidance* Factors" and the *"Targeted Actions."*

Guidance Determination Factors

Targeted Actions	(LPA) Leadership Performance Assessment	(LAAI) Leading Leadership Attributes	(LAAI) Lagging Leadership Attributes	(PLA) Project Leadership Assessment	(TJA) Transformation Journey Approach	(TST) Transformation Support Team
1. Finalize and Communicate My Transformation Vision	X	X	X	X	X	X
2. Improve My "Lagging" Leadership Attributes	X		X			
3. Improve Project Leadership Capacity				X		X
4. Establish and Utilize My Transformation Support Team			X		X	X
5. Finalize and Implement My Communications Strategy					X	X
6. Improve My Focus and Use of the 7-Habits of Successful Project Leaders	X			X		
7. Improve My Understanding and Use of Micro-affirmations and my Ability to Address Micro-inequities		X				X
8. Identify and Harness the Wisdom of Career Mentor #1					X	X
9. Identify and Harness the Wisdom of Career Mentor #2					X	X
10. Identify and Harness the Wisdom of Career Mentor #3					X	X

NOTE: If you feel that the suggested guidance is not consistent with your current "mindset" and Leadership perspectives, you should consider revising the responses to the exercise(s) in question, updating the *Transformation Health Insight Summary*, as appropriate, and following the revised path for suggested action planning guidelines.

Action Planning Guidelines

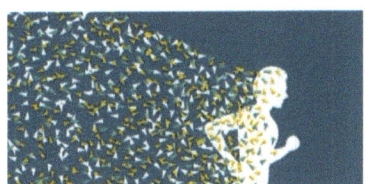

Guideline #1

Based on your **Leadership Performance Assessment Index**, you should understand that you are in the upper half of management in this area. Congratulations. In order to become an Outstanding Leader you need to enhance this strength.

- What areas were your two lowest scores? Focus on these two areas for the next 21 days - concentrate on improving a little each day. This is your first step. Include this as an objective in **Action #2** *(Improving My "Lagging" Leadership Attributes).*
- Ask a trusted colleague what he or she thinks you need to do differently. Notice we said "differently," not better. Sometimes valuable incremental improvement comes from a slight change of approach, not a major overhaul.
- What were your second and third highest scores? Think about how you can effectively use these attributes more in your interactions to improve just a little each day.

Guideline #2

Based on your **Leadership Performance Assessment Index**, you should not be discouraged; this exercise demonstrates a need for improvement, which is what you wanted to learn.

- Find your three lowest scoring Attributes and develop a few objectives in **Action #2** *(Improving My "Lagging" Leadership Attributes)* to elevate your scores. Most likely, this will involve taking a different approach, more thought before action or finding a different way to interact.
- For the next 21 days, make it a priority to track your progress. Incremental, sustainable progress over 21 days will become permanent.
- Keep a success journal - a few notes in a diary (i.e. *"today I looked at the Lockwood project from a CEO's perspective and it validated my ability to think and plan like a Leader."*)

You can do this. You will also benefit greatly from your work here.

Action Planning Guidelines

Guideline #3

Based on your number of **Leading Attributes**, you should embrace your strongest Leadership Attributes and use them as the foundation of your transformation. Because this is an improvement opportunity, approach it by including a few objectives in **Action #3** to achieve the following:

- Select the two lowest Leading Attributes and develop objectives in your Action Plan to improve them.
- Dedicate 14 days for intense focus on those two items - keep a journal of what you did to improve.
- Review the remaining Leading Attributes. What is one thing you can do to improve each of them?
- Dedicate time each day to improve one of these Attributes.

Guideline #4

To create the most effective plan to improve your **Leading Attributes**, you should recognize that you already possess some of the most important skills required of a Leader. With focused effort, you can improve them at the same time you improve the Lagging Attributes. Because this is an improvement opportunity, approach it by including a few objectives in **Action #2 and Action #3** to achieve the following:

- Take a moment to celebrate your strengths and what it took to get to where you are.
- Think about how you can still improve these strengths to advance your career.
- Determine how you can leverage your strengths to improve the Lagging Attributes and include objectives in your Action plan that target this approach.
- Remember that you must still build on these Attributes, not presume they will grow on their own.

Action Planning Guidelines

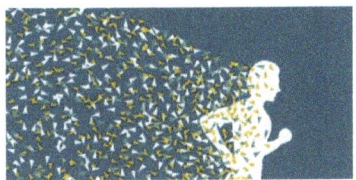

Guideline #5

Based on your number of **Lagging Attributes**, you should remember that you are investing time in this program to advance your career. Through this exercise, you have identified specific attributes to focus your improvement efforts.

The importance of **Action #2** *(Improve My "Lagging" Leadership Attributes)* cannot be minimized, this is the crux of your career advancement, embrace it as such.

- Which two Attributes need the least improvement? Develop objectives in your Action Plan to attack these two first. Again, focus your efforts for the next 21 days and notice the improvement.
- Which attribute needs the most work? For this attribute ask someone close to you (personal or professional) what they think you need to do. Listen and include a 30-day period in the objectives in your Action Plan to track your improvement.
- Do not allow frustration to ruin your improvement. Progress is not always linear, there are frustrating days. You have to work through them to become the Leader you envision.

Guideline #6

To create the most effective plan to improve your **Lagging Attributes**, you should re-read the Transformation Framework associated with your Leading Attributes - note your strengths as well as the areas you are going to improve. Use this as a *"this is what I need to do"* list, not a reason to feel negative.

Action #2 *(Improve My "Lagging" Leadership Attributes)* is your opportunity to say, *"I can and will improve."*

- Which Attribute requires the least time to change? Address it immediately to create an atmosphere of success. Daily progress fuels continual progress.
- Review the longest improvement time and create weekly and monthly milestones in your Action Plan. You need to have victories to celebrate.
- Create a monthly review of your plan and congratulate yourself when you make improvement.

Action Planning Guidelines

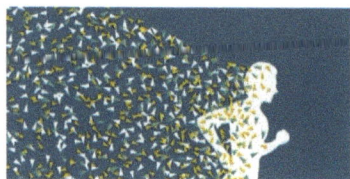

Guideline #7

Based on your **Project Leadership Assessment Index**, you are aligned well with your current organization and your organization's culture and systems.

> ➤ As you develop your objectives for **Action #3** (*Improve Project Leadership Capacity*), you should focus on specific improvement activities that target the areas in the Project Leadership Assessment which have the two lowest scores. An improvement in these areas may make your transformation smoother within your current organization.

> ➤ As you develop your objectives for **Action #6** (*Improve My Focus and Use of the 7-Habits of Successful Project Leaders*), you should determine which of the 7-Habits would benefit you the most at this time, based on your current project and work assignments. Limit your list to 2 or 3 Habits. Then, develop a few measureable short-term and longer-term objectives that would push you toward mastering these habits.

Guideline #8

Your **Project Leadership Assessment Index** indicates that you are in the healthy range, but could still benefit from some focus on alignment.

> ➤ As you develop your objectives for **Action #3** (*Improve Project Leadership Capacity*), you should develop at least one objective targeting each of the five Project Leadership Assessment areas which can help you understand the issue(s) and how you can improve your current alignment. It is important to know what can be changed regarding your needs and what your current organization "can" provide you. This will provide the insight you will need to determine if you may have to seek your new Outstanding Project Leader role in a different organization.

> ➤ As you develop your objectives for **Action #6** (Improve My Focus and Use of the 7-Habits of Successful Project Leaders), you should determine which of the 7-Habits would benefit you the most at this time, based on your current project and work assignments. Limit your list to 2 or 3 Habits. Then, develop a few measureable short-term and longer-term objectives which will push you toward mastering these habits.

Action Planning Guidelines

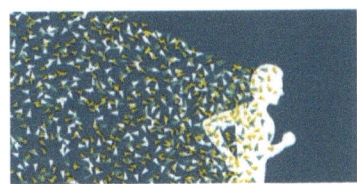

Guideline #9

Based on your **Transformation Journey Approach Index**, it appears that <u>you are</u> satisfied with the work you have done to document how you will approach your transformation journey. Your Index indicates that you have some initial ideas as to how you will communicate your Personal Leadership Vision; how you will put your Support Team in place; and how you will engage with key stakeholders and Mentorship prospects.

- You should go back to the good work you have completed on pages 80 through 84 and make sure your responses are complete and "current" based on where you are at this point in your transformation. Since you first worked on this exercise, you a have covered a lot of material and have learned more about yourself and what will be required of you to complete your transformation journey.
- Once you are finished with your review and possible updates, you should utilize the thoughts you have documented to help you develop some specific objectives that will support your successful completion of **Action #4**, **Action #5** and **Action #8** through **Action #10**.

Guideline #10

Based on your **Transformation Journey Approach Index**, it appears that you <u>are not</u> satisfied with the work you have done to document how you will approach your transformation journey. No worries. This is often the case.

- You should simply go back to the exercise (pages 80 through 84), take the time needed, and provide thoughtful responses to all the questions asked based on where you are at this point in your transformation. Since you first worked on this exercise, you have covered a lot of material and have learned more about yourself and what will be required of you to complete your transformation journey.
- Once you have completed and are satisfied with your responses, utilize these thoughts to help you develop some specific objectives that will support your successful completion of **Action #4**, **Action #5** and **Action #8** through **Action #10**.
- You can do this. If you need more assistance, consider reaching out to a trusted partner or contacting a DUPP Program Coach. A little help at this point will go a long way toward ensuring your transformation success.

Action Planning Guidelines

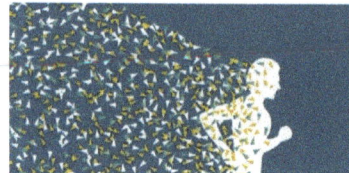

Guideline #11

Great Job! Based on your **Transformation Support Team Index**, <u>you have</u> listed at least one Support Team prospect for each of the six character types. Your Support Team will play an instrumental role in your transformation.

> ➤ You should go back and review the entire discussion on *"Imagining and Constructing Your Transformation Support Team"* (Pages 87 through 94) to make sure you understand the six character types.
> ➤ Then, add additional prospects to your list. Remember, you should use all of your professional networks to identify possible candidates. If you have never met or had an extended conversation with someone who would make a great candidate, get to know them. You can determine their character type later.
> ➤ Keep in mind that your Support Team's make-up, professional background and contacts can help you accomplish the objectives required to achieve **Action #3, Action #4, Action #5** and **Action #7 through Action #10.**

Guideline #12

Based on your **Transformation Support Team Index**, you <u>have not</u> listed at least one Support Team prospect for each of the six character types. Remember, your Support Team will play an instrumental role in your transformation.

> ➤ You should go back and review the entire discussion on *"Imagining and Constructing Your Transformation Support Team"* (Pages 87 through 94) to make sure you understand the six character types.
> ➤ Then, add at least one Support Team prospect for each of the six character types. You should use all of your professional networks to identify possible candidates. If you have never met or had an extended conversation with someone who would make a candidate, get to know them. You can determine their character type later.
> ➤ Keep in mind that your Support Team's make-up, professional background and contacts can help you accomplish the objectives required to achieve **Action #3, Action #4, Action #5** and **Action #7 through Action #10.**

18.

Developing Your Transformation Action Plan

Utilizing the guidance provided in the section 17 and the transformational "thoughts" you have formulated over the time it has taken you to get to this point in the DUPP Program, it is now time to develop an appropriate set of objectives. As in any planning process, setting and accomplishing the proper set of objectives will result in increasing the probability of you executing the actions required to ultimately reach your goal.

In addition to the thought and planning required to accomplish your objectives, the need to *harvest* the strategic conversations and professional support necessary to propel your transformation in the proper direction is also important. Accomplishing this feat is really a *humanistic* art that requires both a personal touch and creativity.

The art of harvesting conversations and professional support that matters is also known as *Participatory Leadership*. It is a style of Leadership that scales up from the *personal* to the *systemic* using personal practice, dialogue, facilitation and the co-creation of innovation to address complex challenges, such as personal and professional transformations. It uses a range of powerful methodologies to harness the collective wisdom of others by:

- Engaging with people where they are;
- Creating spaces for listening; and
- Taking advantage of opportunities to build relationships.

Participatory Leadership is an ongoing practice of all Outstanding Project Leaders. It demands that you embrace your strongest Leadership attributes and consciously deliver your best. You should reach out for any additional coaching assistance to learn how to effectively use this Leadership style while developing and executing your Transformation Action Plan.

Okay. Take your time, keep your focus and develop a great plan. Your transformation is just around the corner.

TRANSFORMING FROM PROJECT MANAGER TO PROJECT LEADER

TRANSFORMATION ACTION PLANNING

The following pages contain your Transformation Action Planning outline. It is comprised of a set of targeted actions, which establishes your plan's baseline. It also directs your focus on what is required of you to achieve your ultimate goal of becoming an Outstanding Project Leader.

The Transformation Action Planning process involves following the suggested "Guidelines" [as set forth in Section 17] and thoughtfully developing a set of specific objectives and milestones for each action in the outline.

Successfully executing your Transformation Action Plan should be your primary developmental focus during the remainder of your transformation journey. The final Sections of this workbook will provide some direction on *finalizing, executing* and *tracking* your plan.

Action #1: Finalize and Communicate My Transformation Vision

Specific Objective	Measurement/Milestones	Attainable?	Relevance?	Time Frame?	Achieved? Y/N
Example: Define and begin an exercise program.	Track duration and frequency of exercise program.	Yes, need to check with my doctor to ensure my plan is appropriate. Adjust plan as advised.	Supports goal to become physically fit.	First month, work up to steady routine. Maintain every month thereafter.	
1. Finalize my *Personal Leadership Vision* drafted in the DUPP Workbook (Page 63).					
2. Communicate my *Personal Leadership Vision* to my friends and family members.					
3. Communicate my Vision to the appropriate co-workers and team members.					
4. Communicate my Vision to the appropriate superiors.					
5. Communicate my Vision to my Support Team Members.					
6. Include my "Personal and Professional Affirmations" in my daily self-talk.					

Action #2: Improve My "Lagging" Leadership Attributes

Specific Objective	Measurement/Milestones	Attainable?	Relevance?	Time Frame?	Achieved? Y/N
Example: Define and begin an exercise program.	Track duration and frequency of exercise program.	Yes, need to check with my doctor to ensure my plan is appropriate. Adjust plan as advised.	Supports goal to become physically fit.	First month, work up to steady routine. Maintain every month thereafter.	

Action #3: Improve Project Leadership Capacity

Specific Objective	Measurement/Milestones	Attainable?	Relevance?	Time Frame?	Achieved? Y/N
Example: Define and begin an exercise program.	Track duration and frequency of exercise program.	Yes, need to check with my doctor to ensure my plan is appropriate. Adjust plan as advised.	Supports goal to become physically fit.	First month, work up to steady routine. Maintain every month thereafter.	

Action #4: Establish and Utilize My Transformation Support Team

Specific Objective	Measurement/Milestones	Attainable?	Relevance?	Time Frame?	Achieved? Y/N
Example: Define and begin an exercise program.	Track duration and frequency of exercise program.	Yes, need to check with my doctor to ensure my plan is appropriate. Adjust plan as advised.	Supports goal to become physically fit.	First month, work up to steady routine. Maintain every month thereafter.	

Action #5: Finalize and Implement My Communications Strategy

Specific Objective	Measurement/Milestones	Attainable?	Relevance?	Time Frame?	Achieved? Y/N
Example: Define and begin an exercise program.	Track duration and frequency of exercise program.	Yes, need to check with my doctor to ensure my plan is appropriate. Adjust plan as advised.	Supports goal to become physically fit.	First month, work up to steady routine. Maintain every month thereafter.	

Action #6: Improve My Focus and Use of the 7-Habits of Successful Project Leaders

Specific Objective	Measurement/Milestones	Attainable?	Relevance?	Time Frame?	Achieved? Y/N
Example: Define and begin an exercise program.	Track duration and frequency of exercise program.	Yes, need to check with my doctor to ensure my plan is appropriate. Adjust plan as advised.	Supports goal to become physically fit.	First month, work up to steady routine. Maintain every month thereafter.	

Action #7: Improve My Understanding and Use of Micro-affirmations and My Ability to Address Micro-inequities

Specific Objective	Measurement/Milestones	Attainable?	Relevance?	Time Frame?	Achieved? Y/N
Example: Define and begin an exercise program.	Track duration and frequency of exercise program.	Yes, need to check with my doctor to ensure my plan is appropriate. Adjust plan as advised.	Supports goal to become physically fit.	First month, work up to steady routine. Maintain every month thereafter.	

Driving Ultimate Project Performance: Transforming from Project Manager to Project Leader

Action #8: Identify and Harness the Wisdom of Career Mentor #1

Specific Objective	Measurement/Milestones	Attainable?	Relevance?	Time Frame?	Achieved? Y/N
Example: Define and begin an exercise program.	Track duration and frequency of exercise program.	Yes, need to check with my doctor to ensure my plan is appropriate. Adjust plan as advised.	Supports goal to become physically fit.	First month, work up to steady routine. Maintain every month thereafter.	

Action #9: Identify and Harness the Wisdom of Career Mentor #2

Specific Objective	Measurement/Milestones	Attainable?	Relevance?	Time Frame?	Achieve? Y/N
Example: Define and begin an exercise program.	Track duration and frequency of exercise program.	Yes, need to check with my doctor to ensure my plan is appropriate. Adjust plan as advised.	Supports goal to become physically fit.	First month, work up to steady routine. Maintain every month thereafter.	

Action #10: Identify and Harness the Wisdom of Career Mentor #3

Specific Objective	Measurement/Milestones	Attainable?	Relevance?	Time Frame?	Achieve? Y/N
Example: Define and begin an exercise program.	Track duration and frequency of exercise program.	Yes, need to check with my doctor to ensure my plan is appropriate. Adjust plan as advised.	Supports goal to become physically fit.	First month, work up to steady routine. Maintain every month thereafter.	

Finalizing and Executing Your Action Plan

Executing your action plan is an important part of your transformation process. Any professional athlete will tell you that the final moments before a game or event are excruciating – they just want to get started. We hope you feel that way also.

Are you ready to complete your Transformation from "Manager" to "Leader"?

Well, what happens next is exclusively within your realm of responsibility.

In this section we have included an *Action Plan Launch Guide.* The guide is a list of important items you should keep in mind as you finalize and launch your Transformation Action Plan. A thorough and thoughtful review of the entire plan will help ensure that it is complete and ready to be executed.

We have also included an *Action Plan Execution Guide.* In this guide you will find a description of the *"three components of execution"* and a specific set of *behaviors* and *techniques* which support each component.

You should keep all of this additional guidance "top-of-mind" as you move forward, execute your plan and reach your goal.

Remember, when you execute today, you win tomorrow!

> *"Writing in a journal reminds you of your goals and of your learning in life. It offers a place where you can hold a deliberate, thoughtful conversation with yourself."*
>
> — Robin S. Sharma

Action Plan Launch Guide

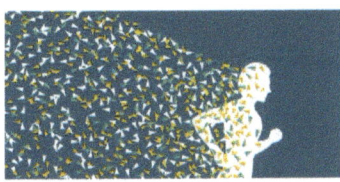

The following are items you should keep in mind as you finalize and launch your Transformation Action Plan.

- **Your plan does not have to be perfect.** However your plan does need to cover the Leadership Attributes you chose to invest time and energy to improve.

- **Your plan needs to be fluid.** You might project four months to improve your communication skills yet accomplish your goal in two months. Conversely, you may think you need two weeks to select a mentor but need five. Things happen – Outstanding Leaders are flexible in such matters.

- **Overcome your fear of failure.** To improve your career arc, you need to move from internal negative talk and move toward your goal.

- **Trust your instincts.** If you have been honest in your assessments, you know the improvements you need to make. Believe in you and move forward.

- **Commit to self-improvement.** The plan that you have developed is a roadmap to success. If you are willing to invest the time and effort to succeed, you will.

- **Get started.** Do not wait for the ideal time or situation, do something today!

- **Make it part of your day – schedule it.** Schedule 30 minutes for self-improvement into your day; you can invest in yourself anywhere - at home, at work, at lunch or the gym. It does not matter when or where, what matters is doing it.

- **Remember, you are responsible for executing your plan.** However, partner with someone who cares about your success and will help hold you accountable. This could be - your mentor, spouse, friend or coworker. Make it a weekly task to provide an update to him or her. The best way keep your focus is to keep a diary, where you enter your progress daily or weekly. You should faithfully utilize the *Monthly Tracking Templates* included in section 20 of this workbook.

- **Expect plateaus and setbacks.** Progress is rarely linear; there are highs and lows - progress and regression. Work through it.

Action Plan Launch Guide

- **Look back.** Every month, review your progress honestly. What did I accomplish? Am I on track?
- **Celebrate milestones and achievements.** If someone compliments you for demonstrating a Leadership skill, celebrate that achievement. Record successes in your diary and watch the list grow.
- **Seek feedback.** Ask your mentor for his or her opinion of your progress. Listen carefully to both the positive and negative statements. Then, incorporate that information into your plan.
- **Seek independent confirmation.** Ask your Mentor or coach for an independent review of your progress.
- **Never give up.** This is your investment in you and your future; it is one of the most important investments you can make. It is worth the struggles.

Do you have more thoughts you desire to keep in mind as you launch your Action Plan?

Action Plan Execution Guide

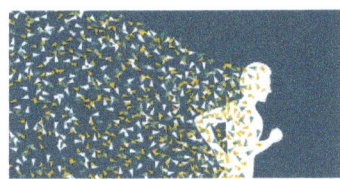

Transformation Action Plan execution is an easy concept to talk about, but it is a difficult, life changing endeavor. The major concern is that it is a real challenge to measure and manage multiple self-development objectives at the same time you are maintaining your focus on your job and your family. However, once you compartmentalize and prioritize the daily activities and behaviors associated with *your professional development*, *your job* and *your family*, you have something that you can measure and something you can manage.

To help you with *your professional development*, we have detailed below the basics of three components of execution and the specific behaviors associated with each. When you consistently *sharpen your focus*, *build your competence* and *ignite your passion*, you plant the seeds for a transformation victory. Initially, you may not see tangible results, but rest assured professional growth and change is occurring under the surface. As you stick with it, momentum builds, creating a clear view of how you will accomplish your vision.

Components of Execution	Specific Behaviors
1. Sharpen your Focus. Focus provides the clarity necessary to make decisions that support your most important objectives and actions. It results in a clearly-defined pathway to transformational success. A sharp focus answers the "what" question: *What do I need to do to execute my objectives and actions?*	**Keep it simple:** ➤ Think in threes to simplify objective, metrics, and actions. ➤ Apply the 80/20 principle to focus on the 20 percent that are your vital few "things-to-do" this week or this month. **Identify your one thing:** ➤ Decide what is most important to do this week or this month--the one activity that most directly helps you execute your plan. ➤ Ask yourself, "What is the most valuable thing I can do right now?"

Components of Execution	Specific Behaviors
	Know when to say no: ➤ Create some quality time to filter new opportunities to develop your Leadership skills or arrange a meeting with a Mentor. ➤ Say no to activities, meetings, and projects that do not directly support your Action Plan and Leadership growth.
2. Build your Competence. Competence is used here in the broadest sense of the term. It encompasses all the skills, systems, processes and tools you can use to achieve your objectives and actions. The result is the ability to commit to, measure and hit your targets. Building competence answers the "how" question: *How will I execute my Transformation Plan?*	**Treasure your talent:** ➤ Select the best possible candidates to ask to join your Support Team or to become a professional Mentor. ➤ Treat improving your Leadership skills and presence as a perpetual priority. ➤ Communicate for success with four steps: explain, ask, involve, and appreciate. **Get systematic:** ➤ Create repeatable systems and practices as a means of sharing your *"Personal Transformation Vision"*. ➤ Collaborate with clear roles. (Lead the discussions, Do the homework, Share your expertise, Get informed, Listen, Learn) ➤ Innovate daily – try new ways to improve and to *"think like a Leader."* **Balance your perspective:** ➤ Seek both dashboard (general) and under-the-hood (detailed) knowledge about your transformation progress. ➤ Track Leading and Lagging Attribute use and improvement.

Components of Execution	Specific Behaviors
3. Ignite your Passion. Passion creates a sense of connectedness. It builds a connection between your Senior Management, your Peers, your Support Team Members and your Mentors. Passion also builds a connection to our human need for meaningful improvement and professional growth as well as a connection to each individual's sense of value and contribution. Igniting passion answers the "why" question: *Why am I executing my Transformation Plan?*	**Paint the picture:** ➢ Connect each planned objective to the broader action and purpose. ➢ Always answer the fundamental four questions: What? Why? When? Where? ➢ Communicate proactively to avoid misunderstandings surrounding your transformation journey and the *"silence spiral."* **Give what you want:** ➢ Show uncommon respect with common courtesy. ➢ Openly appreciate support and advice, as well as the person behind it. ➢ Encourage yourself with daily affirmations and remember the three simple words: Believe in Yourself. **Create connections:** ➢ Use meaningful rituals. ➢ Be accessible to your Support Team, your Mentors and Professional Coaches. ➢ Be who you are and nobody else.

"Human behavior flows from three main sources: desire, emotion, and knowledge."

— **Plato**

20.

Transformation Action Tracking

Congratulations. To arrive at this point means that you have successfully completed all of the workbook exercises to your satisfaction and taken advantage of the insights and developmental opportunities it presents. It also indicates that you are ready to continue to move forward and complete your transformation from a good *Project Manager* to an *Outstanding Project Leader*.

As we shared in the introduction, "Everyone is one good decision away from a better life" and *you* are one-step closer because you made the decision to participate in the *Driving Ultimate Project Performance (DUPP) Program* and to take advantage of the self-paced *DUPP Workbook and Coaching Guide.* Your decision to take action to improve your Project Leadership skills will create a more fulfilling and impactful life.

What is next?

It is now time to focus on making sure that you are successful in completing your transformation by tracking your progress, at least on a monthly basis.

How can tracking your progress help you execute your plan?

An effective method of tracking transformational success can help you stay focused. Many smart professionals who we have had the opportunity to meet have failed to accomplish significant career changes that they so deeply desired, not because they lacked the skills but because they consistently would lose sight of what they really wanted to achieve.

Tracking helps you to focus on the important things you need to do. You have just documented in a comprehensive plan what you need to do in order to move forward and achieve your ultimate goal. If you do not properly track your progress during the plan's execution, you are more likely to focus on your failures and not your successes. It is natural for any of us to focus on the negative side of things. Even if you accomplished something last month, mishaps you may experience next month can make you forget your achievements.

To avoid dwelling on negative things, you should make it a habit to track the actions you have taken and the achievements you have made to date.

Our most successful clients often share with us that after a few months into executing their Transformation Plan, they began to notice a pattern, which allowed them to concentrate on the most important objectives. This pattern provided clarity and, as a result, they began to more quickly achieve their objectives and move toward their ultimate goals.

To assist you in your Transformation Action tracking effort, we have included in this section a set of *Monthly Transformation Action Tracking* templates.

It is important to track which objectives in your plan that you successfully complete each month and focus on the objectives scheduled to be "tackled" the following month.

Also, set aside some quality time each month for some introspection regarding your transformation progress. Attempt to honestly reflect on how you feel about your transformation journey and what challenges you should prioritize based on your progress to date. Do not hesitate to share your thoughts and feelings with appropriate Support Team Members, Mentors or Professional Coaches.

The key to all transformation success is persistence, motivation and hard work. As Napoleon Hill once put it, *"Patience, persistence and perspiration make an unbeatable combination for success."*

Monthly Transformation Action Tracking

Month _____

	Objectives Completed This Month	Objectives Planned for Next Month	What Feels Good About Your Progress?	What Are your Challenges?
Action #1 Finalize and Communicate My Transformation Vision				
Action #2 Improve My Lagging Leadership Attributes				
Action #3 Improve Project Leadership Capacity				
Action #4 Establish and Utilize My Transformation Support Team				
Action #5 Finalize and Implement My Communications Strategy				

	Objectives Completed This Month	Objectives Planned for Next Month	What Feels Good About Your Progess?	What Are your Challenges?
Action #6 Improve My Focus and Use of the 7-Habits of Successful Project Leaders	<ul><li></li><li></li><li></li></ul>	<ul><li></li><li></li><li></li></ul>	<ul><li></li><li></li><li></li></ul>	<ul><li></li><li></li><li></li></ul>
Action #7 Improve My Understanding and Use of Micro-affirmations and My Ability to Address Micro-inequities	<ul><li></li><li></li><li></li></ul>	<ul><li></li><li></li><li></li></ul>	<ul><li></li><li></li><li></li></ul>	<ul><li></li><li></li><li></li></ul>
Action #8 Identify and Harness the Wisdom of Career Mentor #1	<ul><li></li><li></li><li></li></ul>	<ul><li></li><li></li><li></li></ul>	<ul><li></li><li></li><li></li></ul>	<ul><li></li><li></li><li></li></ul>
Action #9 Identify and Harness the Wisdom of Career Mentor #2	<ul><li></li><li></li><li></li></ul>	<ul><li></li><li></li><li></li></ul>	<ul><li></li><li></li><li></li></ul>	<ul><li></li><li></li><li></li></ul>
Action #10 Identify and Harness the Wisdom of Career Mentor #3	<ul><li></li><li></li><li></li></ul>	<ul><li></li><li></li><li></li></ul>	<ul><li></li><li></li><li></li></ul>	<ul><li></li><li></li><li></li></ul>

Monthly Transformation Action Tracking

Month _____

	Objectives Completed This Month	Objectives Planned for Next Month	What Feels Good About Your Progress?	What Are your Challenges?
Action #1 Finalize and Communicate My Transformation Vision				
Action #2 Improve My Lagging Leadership Attributes				
Action #3 Improve Project Leadership Capacity				
Action #4 Establish and Utilize My Transformation Support Team				
Action #5 Finalize and Implement My Communications Strategy				

	Objectives Completed This Month	Objectives Planned for Next Month	What Feels Good About Your Progess?	What Are your Challenges?
Action #6 Improve My Focus and Use of the 7-Habits of Successful Project Leaders				
Action #7 Improve My Understanding and Use of Micro-affirmations and My Ability to Address Micro-inequities				
Action #8 Identify and Harness the Wisdom of Career Mentor #1				
Action #9 Identify and Harness the Wisdom of Career Mentor #2				
Action #10 Identify and Harness the Wisdom of Career Mentor #3				

Monthly Transformation Action Tracking

Month _____

	Objectives Completed This Month	Objectives Planned for Next Month	What Feels Good About Your Progress?	What Are your Challenges?
Action #1 Finalize and Communicate My Transformation Vision				
Action #2 Improve My Lagging Leadership Attributes				
Action #3 Improve Project Leadership Capacity				
Action #4 Establish and Utilize My Transformation Support Team				
Action #5 Finalize and Implement My Communications Strategy				

	Objectives Completed This Month	Objectives Planned for Next Month	What Feels Good About Your Progress?	What Are your Challenges?
Action #6 Improve My Focus and Use of the 7-Habits of Successful Project Leaders				
Action #7 Improve My Understanding and Use of Micro-affirmations and My Ability to Address Micro-inequities				
Action #8 Identify and Harness the Wisdom of Career Mentor #1				
Action #9 Identify and Harness the Wisdom of Career Mentor #2				
Action #10 Identify and Harness the Wisdom of Career Mentor #3				

Monthly Transformation Action Tracking

Month _____

	Objectives Completed This Month	Objectives Planned for Next Month	What Feels Good About Your Progress?	What Are your Challenges?
Action #1 Finalize and Communicate My Transformation Vision				
Action #2 Improve My Lagging Leadership Attributes				
Action #3 Improve Project Leadership Capacity				
Action #4 Establish and Utilize My Transformation Support Team				
Action #5 Finalize and Implement My Communications Strategy				

	Objectives Completed This Month	Objectives Planned for Next Month	What Feels Good About Your Progress?	What Are your Challenges?
Action #6 Improve My Focus and Use of the 7-Habits of Successful Project Leaders	▪ ▪ ▪	▪ ▪ ▪	▪ ▪ ▪	▪ ▪ ▪
Action #7 Improve My Understanding and Use of Micro-affirmations and My Ability to Address Micro-inequities	▪ ▪ ▪	▪ ▪ ▪	▪ ▪ ▪	▪ ▪ ▪
Action #8 Identify and Harness the Wisdom of Career Mentor #1	▪ ▪ ▪	▪ ▪ ▪	▪ ▪ ▪	▪ ▪ ▪
Action #9 Identify and Harness the Wisdom of Career Mentor #2	▪ ▪ ▪	▪ ▪ ▪	▪ ▪ ▪	▪ ▪ ▪
Action #10 Identify and Harness the Wisdom of Career Mentor #3	▪ ▪ ▪	▪ ▪ ▪	▪ ▪ ▪	▪ ▪ ▪

Monthly Transformation Action Tracking

Month _____

	Objectives Completed This Month	Objectives Planned for Next Month	What Feels Good About Your Progress?	What Are your Challenges?
Action #1 Finalize and Communicate My Transformation Vision				
Action #2 Improve My Lagging Leadership Attributes				
Action #3 Improve Project Leadership Capacity				
Action #4 Establish and Utilize My Transformation Support Team				
Action #5 Finalize and Implement My Communications Strategy				

	Objectives Completed This Month	Objectives Planned for Next Month	What Feels Good About Your Progess?	What Are your Challenges?
Action #6 Improve My Focus and Use of the 7-Habits of Successful Project Leaders	▪ ▪ ▪	▪ ▪ ▪	▪ ▪ ▪	▪ ▪ ▪ ▪
Action #7 Improve My Understanding and Use of Micro-affirmations and My Ability to Address Micro-inequities	▪ ▪ ▪	▪ ▪ ▪	▪ ▪ ▪	▪ ▪ ▪ ▪
Action #8 Identify and Harness the Wisdom of Career Mentor #1	▪ ▪ ▪	▪ ▪ ▪	▪ ▪ ▪	▪ ▪ ▪ ▪
Action #9 Identify and Harness the Wisdom of Career Mentor #2	▪ ▪ ▪	▪ ▪ ▪	▪ ▪ ▪	▪ ▪ ▪ ▪
Action #10 Identify and Harness the Wisdom of Career Mentor #3	▪ ▪ ▪	▪ ▪ ▪	▪ ▪ ▪	▪ ▪ ▪ ▪

Monthly Transformation Action Tracking

Month_____

	Objectives Completed This Month	Objectives Planned for Next Month	What Feels Good About Your Progress?	What Are your Challenges?
Action #1 Finalize and Communicate My Transformation Vision				
Action #2 Improve My Lagging Leadership Attributes				
Action #3 Improve Project Leadership Capacity				
Action #4 Establish and Utilize My Transformation Support Team				
Action #5 Finalize and Implement My Communications Strategy				

	Objectives Completed This Month	Objectives Planned for Next Month	What Feels Good About Your Progess?	What Are your Challenges?
Action #6 Improve My Focus and Use of the 7-Habits of Successful Project Leaders	▪ ▪ ▪	▪ ▪ ▪ ▪ ▪	▪ ▪ ▪	▪ ▪ ▪ ▪
Action #7 Improve My Understanding and Use of Micro-affirmations and My Ability to Address Micro-inequities	▪ ▪ ▪	▪ ▪ ▪ ▪ ▪	▪ ▪ ▪	▪ ▪ ▪ ▪
Action #8 Identify and Harness the Wisdom of Career Mentor #1	▪ ▪ ▪	▪ ▪ ▪ ▪ ▪	▪ ▪ ▪	▪ ▪ ▪ ▪
Action #9 Identify and Harness the Wisdom of Career Mentor #2	▪ ▪ ▪	▪ ▪ ▪ ▪ ▪	▪ ▪ ▪	▪ ▪ ▪ ▪
Action #10 Identify and Harness the Wisdom of Career Mentor #3	▪ ▪ ▪	▪ ▪ ▪ ▪ ▪	▪ ▪ ▪	▪ ▪ ▪ ▪

LIST OF EXERCISES

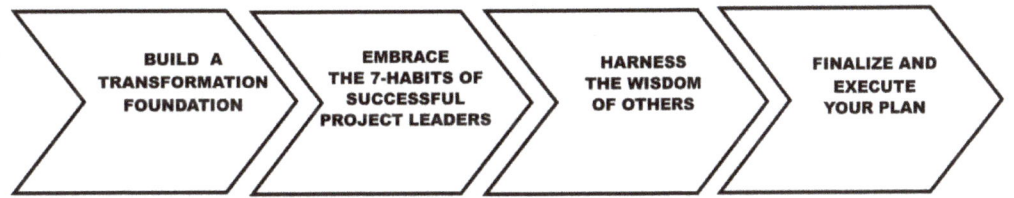

	Exercises	Page
1)	Leadership Performance Assessment	15
2)	Determine Your Leadership Attribute Gaps	57
3)	Identify Your "Lagging" and "Leading" Attributes	58
4)	Develop Your Lagging Attribute Improvement Plan	61
5)	Construct Your Personal Leadership Vision	63
6)	List Your Transformation Affirmations	64
7)	Project Leadership Assessment - Current Organizational Alignment	69
8)	Project Leadership Assessment - Leadership Type	71
9)	Project Leadership Assessment - Leadership Behaviors	73
10)	Project Leadership Assessment - Leadership Resilience	75
11)	Project Leadership Assessment - Leadership Development Perspective	77
12)	Planning Your Transformation Journey	79
13)	Imagining and Constructing Your Transformation Support Team	87
14)	Outlining Your Transformation Team Communications Strategy	95
15)	CASE STUDY: I. Communicating Vertically, Horizontally and Often	106
16)	CASE STUDY: II. Planning as a Team, Executing as a Team, Winning as a Team	110
17)	CASE STUDY: III. Proactively and Fearlessly Managing Project and Resource Change	114
18)	CASE STUDY: IV. Linking Risk to Common Objectives with a Compelling Vision	118
19)	CASE STUDY: V. Approaching Problem Solving as a Creative and Analytical Process	122
20)	CASE STUDY: VI. Maintaining Project Quality with Incremental Measures and Team Focus	126
21)	CASE STUDY: VII. Accepting Your Role as the Chief Confidence Builder	130
22)	Manager Versus Leader	139
23)	Learning How to "Think Like a Leader"	141
24)	Micro-inequities and Micro-affirmation - Questions to Consider	152
25)	Your Career Critical Needs Today	156
26)	Developing an Mentor "Ask" Format	163
27)	Constructing an Mentor "Ask" Message	165
28)	Constructing Your Mentor Conversation Starters	166
29)	Transformation Health Insight Summary (THIS) Development	175
30)	Determining Your Action Planning Guidelines	183

A Project Leader is a Project Manager, but a Project Manager is not necessarily a Project Leader.

ABOUT THE AUTHORS

Jim Grigsby, CRCE, CHC

Jim Grigsby has over 20 years of healthcare financial management experience. He founded Jim Grigsby Consulting in 2006 to implement solutions that improve revenue cycle efficiency and results.

Jim climbed the healthcare finance management ladder from Collector to Director, Patient Financial Services with responsibility for the entire Revenue Cycle: Central Scheduling, Verification, Access, Financial Counseling, Billing, and Collections. As a consultant, Jim helped establish a MSO, manage Access and Cash Improvement Projects and lead numerous process re-engineering and analysis projects.

Jim is also the author of several self-improvement books, including his bestselling books *"The Official Leadership Checklist and Diary for Project Management Professionals"*, *"Don't Tick off the Gators! Managing Problems Before Problems Manage You"* and *"Are You Surrounded By Jerks? How to Deal with the Most Annoying People."*

Ervin (Earl) Cobb, MSEE

Since 2010, Earl Cobb has been the CEO and Managing Partner of Richer Life, LLC — a digital media, trade book publishing and professional services company headquartered in Phoenix, Arizona.

Earl spent the previous 34 years of his career in Fortune 100, Mid-Market and Venture companies as a Systems Engineer, Program/Project Manager, Management Executive and Technology Executive. Earl has held senior management positions with Motorola, The Reynolds & Reynolds Company and Wells Fargo Bank. Earl is the former President, COO and CEO of the high-tech start-up, MedContrax.

Earl is the author of seven other published books, including his bestselling books, *"The Official Leadership Checklist and Diary for Project Management Professionals"*, *"Focused Leadership: What You Can Do Today To Become A More Effective Leader"* and *"The Leadership Advantage: Do More. Lead More. Earn More."*

"Think of The Official Leadership Checklist and Diary as your personal brain catalyst – a handy guide containing thought provoking reminders to keep the leadership aspects of your project on track."

"The book is the consummate guide for PMP professionals to employ and deploy successful project leadership strategies. Most publications focus on the technical aspects of project management while giving little attention to the leadership aspects. This practical guide is a thought provoking resource for project leaders that will enhance project outcomes."~ **Anthony V. Junior, Ph.D., MBA, PMP, Principal, Strategic Consulting Network, LLC**

"This book concept to discuss "Leadership" for PM roles is great. This topic is always relevant." ~ **Deanna Hawkins, PMP, CEO, 7LampsFX**

"Earl Cobb, well-known for helpful, concise business leadership books, has done it again. This time Earl has turned his attention to the world of Project Management, successfully showing PMP's good leadership principles." ~ **Doug Russell, PMP, Financial Counselor and Coach**

"This book is a high-value quick read, packed with useful information and tips. The advice provided will enhance leadership and project management skills of a seasoned professional or one with the responsibility of a first project." ~ **Dr. Kenneth Morton, Infinity Leadership Consulting, LLC**

"The book offers a thoughtful, systematic way to ensure that project managers actively lead teams to bring out the best in teams." ~ **Yvette DeBois, MD MPH**

"The Official Leadership Checklist and Diary is an excellent resource for leaders at all levels, including project leaders. The book's excellent, highly practical content is simple, straightforward, based on the real world and forever relevant. A great personal, professional and organizational investment" ~ **Dan Nielsen, Retired Healthcare CEO, Founder/CEO/Publisher of America's Healthcare Leaders**

"Over my career in Healthcare Financial Management, I have led or managed hundreds of large and complex projects, sometimes with the help of Jim Grigsby. I have always found Jim to be the most organized person on the project. Now I know why! I found the Leadership Checklist and Diary to be right on the mark! I applied some of the principles on a current project as I was reading. The 7 Habits sizzle with common sense. This book has earned a permanent place in my 'toolbox'." ~ **John Midolo Managing Partner RCM Strategies**

Now Available Wherever Quality Books Are Sold

www.ingramcontent.com/pod-product-compliance
Lightning Source LLC
Chambersburg PA
CBHW061140230426

43663CB00027B/2981